Seaman's Guide to Human Factors, Leadership, and Personnel Management

Seaman's Guide to Human Factors, Leadership, and Personnel Management

Authored by
Jose Rodriguez Cordon

CRC Press
Taylor & Francis Group
Boca Raton London New York

CRC Press is an imprint of the
Taylor & Francis Group, an **informa** business

CRC Press
Taylor & Francis Group
6000 Broken Sound Parkway NW, Suite 300
Boca Raton, FL 33487-2742

International Standard Book Number-13 978-0-367-19749-0 (Hardback)

Library of Congress Cataloging-in-Publication Data

Names: Cordon, Jose Rodriguez, 1966- author.
Title: Seaman's guide to human factors, leadership, and personnel management /by Jose Rodriguez Cordon.
Description: Boca Raton, FL: CRC Press/Taylor & Francis Group, 2019. | Includes bibliographical references and index.
Identifiers: LCCN 2019002490| ISBN 9780367197490 (hardback : acid-free paper) | ISBN 9780429243059 (ebook)
Subjects: LCSH: Merchant marine--Manning of vessels. | Merchant marine--Officers--Training of. | Merchant marine--Personnel management. | Leadership--Handbooks, manuals, etc.
Classification: LCC VK531 .C67 2019 | DDC 387.5068/3--dc23
LC record available at https://lccn.loc.gov/2019002490

Visit the Taylor & Francis Web site at
http://www.taylorandfrancis.com

and the CRC Press Web site at
http://www.crcpress.com

To my family. Nothing is possible without you.

Contents

Foreword xi
Preface xiii
Author xv

Chapter 1   Introduction 1

           References 3

Chapter 2   Leadership 5

    2.1   Leadership on Board 5
          2.1.1   Definitions of Leadership 7
          2.1.2   Focuses Centered on the Leader, Trait, or Process? 8
          2.1.3   The Trait Approach 8
          2.1.4   The Leadership Styles Approach 9
          2.1.5   Contingency Models 9
                  2.1.5.1   Fielder's Model 10
                  2.1.5.2   Hersey and Blanchard Situational
                               Leadership Model 10
                  2.1.5.3   Model of Leader Participation 13
                  2.1.5.4   Path to the Goal Model 14
          2.1.6   Leadership Models Focused on Followers:
                 Implicit Leadership Theories 18
          2.1.7   Leadership as a Process of Influence:
                 Idiosyncratic Credit Theory 19
          2.1.8   Charismatic Leadership 20
          2.1.9   Transformational Leadership and Transactional
                 Leadership 21
                  2.1.9.1   Transformational Leadership 22
          2.1.10 Leadership and Intergroup Relations 22
          2.1.11 Intergroup Phenomena 23
          2.1.12 Intergroup Hostility 24
          2.1.13 Intergroup Harmony 25
          2.1.14 Leadership and Organizational Climate 26
          2.1.15 Leadership and Job Satisfaction 26
          2.1.16 Cross-Cultural Leadership 27
          2.1.17 Leadership and Gender 28
          2.1.18 Ethical Leadership and Authentic Leadership 28
    2.2   The Evil Leadership: The Case of *Maria M* 29
          2.2.1   The Crew 30
          2.2.2   The Events 30
          2.2.3   Human Factor 30

2.3   Leadership in Emergency Situations 31
       2.3.1   Decision-Making Under Pressure 32
                2.3.1.1   Role of the Cerebral Hemispheres in
                           Emergency Situations 32
       2.3.2   Fear: Psychological Reaction to an Emergency
                Situation...33
                2.3.2.1   States of Fear and Forms of Appearance......33
                2.3.2.2   Resilience (Ability to Overcome
                           Adversity)...34
       2.3.3   Reactions in Emergency Situations 36
       2.3.4   Verbal Communication in Emergency Situations 36
       2.3.5   Crew Training for Crowd Control 37
                2.3.5.1   Chaos Is Inherent in All Emergencies........37
                2.3.5.2   Training for Crowd Control 37
                2.3.5.3   Special Emphasis on Information
                           for Officers...38
       2.3.6   How to Inform Passengers in Case of Emergency 38
                2.3.6.1   Alarm Signals Are Not Enough.................39
                2.3.6.2   Responsibility Increases the Potential
                           for Action..39
                2.3.6.3   Authority Is a Prerequisite in Crowd
                           Control 40
2.4   Catastrophe of the *Estonia* 40
       2.4.1   The Crew 41
       2.4.2   The Events...41
       2.4.3   The Human Element 42
References 42

Chapter 3   Teamwork...47

3.1   Group Structure..47
       3.1.1   Effects of Group Size 47
       3.1.2   Effects of Group Diversity 48
       3.1.3   Effects of the Combination of Individuals in
                the Group...48
       3.1.4   The Concept of Group Structure...............................48
                3.1.4.1   Intragroup Differentiation 48
                3.1.4.2   Hierarchies in Groups...............................49
       3.1.5   Group Regulations...49
                3.1.5.1   Effects of Deviation with Respect to
                           the Group Norms 50
                3.1.5.2   The "Black Sheep" Effect........................50
       3.1.6   Group Roles..50
       3.1.7   Processes of Influence Inside Groups 51
                3.1.7.1   Conformism 51
                3.1.7.2   Uncertainty and Consensus 52
                3.1.7.3   Factors That Influence Conformism..........52

3.1.8   Minority Influence...53
        3.1.8.1   Mechanisms of Defense against
                  Minority Influence......................................53
3.2   Productivity of the Group...54
      3.2.1   Group Performance...54
      3.2.2   Effects of the Public and Coercion on Productivity......55
               3.2.2.1   Effects of Coordination and
                         Motivation..55
3.3   Group Decision Processes: How Groups Make Decisions......56
      3.3.1   Group Polarization...57
      3.3.2   Biases and Limitations in Group Decision-Making.....58
3.4   Group Thinking..59
      3.4.1   Some Highlights of Intergroup Relations..................60
3.5   Benefits of Teamwork..61
      3.5.1   Characteristics of Effective Teams...........................61
      3.5.2   Role of the Leader in Team Building........................61
      3.5.3   Stages in Team Building..62
      3.5.4   Skills of Team Members..62
      3.5.5   Self-Directed Teams...63
               3.5.5.1   Formation of Self-Directed Teams.............64
               3.5.5.2   Types of Self-Directed Teams....................64
               3.5.5.3   Role of the Leader in Self-Directed
                         Teams...65
      3.5.6   Empowerment of Teams...65
3.6   Managing Diversity: Multicultural Crews.................................65
      3.6.1   Diversity and Its Effects on Groups: General
               Investigations..66
      3.6.2   Cultural Dimensions...66
      3.6.3   Multiculturalism in the Maritime Context.................68
               3.6.3.1   Research on Multiculturalism in the
                         Merchant Navy..68
      3.6.4   Current Panorama of the Crews..................................69
      3.6.5   Challenges of Multicultural Crews............................69
               3.6.5.1   Living and Working Conditions.................69
               3.6.5.2   Idiomatic Differences....................................70
      3.6.6   Non-Verbal Language and Cultural Differences.........71
               3.6.6.1   Ambiguous Gestures and Cultural
                         Differences..75
               3.6.6.2   Some Negotiation Rules...............................76
               3.6.6.3   Considerations of Religious Differences.......76
      3.6.7   Gender Differences On Board......................................77
3.7   When All Teamwork Fails: The Case of the *Bow Mariner*......79
      3.7.1   The Crew...79
      3.7.2   The Events..79
      3.7.3   The Human Factor...80
3.8   Conclusions..81
References..81

Chapter 4   The Psychological Challenges of Life on Board 85

    4.1   Shift Work: The Circadian Rhythms 87
          4.1.1   Sleep and Its Disorders.................................. 87
          4.1.2   Fatigue 90
                   4.1.2.1   Factors That Contribute to Fatigue 91
                   4.1.2.2   Effects of Fatigue.. 91
                   4.1.2.3   Strategies to Combat Fatigue On Board 92
          4.1.3   Uprooting 94
          4.1.4   Long Periods on Board Ship 94
          4.1.5   Multi-Ethnic Crews 94
          4.1.6   Workload and Time Pressure 95
          4.1.7   Demands of the Charge.................................. 95
    4.2   Fatigue and Natural Tragedy: The Case of The
        Exxon Valdez.. 95
          4.2.1   The Crew 96
          4.2.2   The Events.. 96
          4.2.3   The Human Factor.. 97
    References 98

Chapter 5   The Case of VTS 101

    5.1   Risk Factors 102
          5.1.1   Shift Work 102
          5.1.2   Cognitive Overload.. 103
          5.1.3   Stress 104
          5.1.4   Work–Life Balance.. 105
    5.2   Protective Factors 105
    5.3   Conclusion 106
    References 107

Index.. 111

Foreword

For many years, all the improvements made in sea transportation have been focused on technological advances and automatization, trying to get the human element out of the equation. There are many studies on human error, how they may happen and how to avoid them. But human beings are not machines; therefore, they cannot be treated as such.

Recently, great concern about human factors has been raised among stakeholders in seafaring, as it is evident that many maritime accidents happen because of human failures. This concern arises in the form of the so-called Manila Amendments to the International Convention on Standards of Training, Certification and Watchkeeping for Seafarers, which introduced for the first time some psychological features into the officers' background.

How to deal with groups coming from different cultures and becoming a leader are the key aspects for every officer on board. Probably these are the most complex and frustrating elements that every seaman has to face; however, there is no specific preparation in the academy's curricula.

In this guide, I tried to break down all the knowledge that every officer will need to work with people, using what psychology can teach us, keeping the necessary depth but adapted to the seaman's needs. I think this is the book I would like to have read when I first became first mate. This was the drive to write it, and I hope officers across the world find it useful.

Preface

A seaman's life is a process of learning. Every time you step onto a new ship you have to deal with new equipment, new radars, new devices, and surely, new work methods and an ever-increasing amount of paperwork! Those things can be learned through technical manuals, or by observing more experienced people. Moreover, the technical skills required to be an officer are usually taught at the academy.

In contrast, the day I was appointed chief mate for the first time, I realized that the toughest thing had nothing to do with ropes, but with people. Managing crews in such a restrictive, highly structured environment is a complex task, and requires not only leadership knowledge, but also it is essential to understand how people behave on board. People don't perform the same way when they are isolated or when they are in groups or even when they find themselves in the middle of a multi-ethnic group. To sum up, an officer has to become a leader, because it is as much a part of their work as it is to understand how a ship is loaded.

When I got my degree in psychology, I focused my work on everything related to human factors in seafaring. I was required to lecture about the so-called Manila Amendments to the International Convention on Standards of Training, Certification and Watchkeeping for Seafarers, so I started gathering information about leadership. However, as a psychologist, I realized that leadership cannot be taught independently, disregarding the ship's context.

Leadership is a fashionable construct, and you can find a myriad of courses, along with a proliferation of *coaching* phenomena. But this approach is oriented to the world of enterprise, which has nothing in common with that aboard a vessel. Also, some particular characteristics of a ship's environment are unique to it: its imposed structure of power, its isolation and multiculturalism, among others. Also, many coaching courses are merely a collection of inspiring ideas, useless for an officer.

To really understand how people behave on board, it is essential to know the challenges that they face, both socially and psychologically. Furthermore, we must comprehend how human groups behave, particularly in the isolated environment of a vessel. Likewise, officers on passenger vessels must know how to deal with crowds in emergency situations. My experience in these situations drove me to write this guide. My experience in Vessel Traffic Services had the same effect.

For two years, I compiled material about these subjects, reading and analyzing all the relevant research on this topic. Finally, I came to an integrated approach to that part of human factors that helps an officer to manage people, to understand their behavior, and, in many ways, to get people to do their best; all of which, in a few words, means to be a leader.

Author

Jose Rodriguez Cordon was born in Algeciras (Spain) in 1966. He lives in Cadiz (Spain) and works for Cadiz Vessel Traffic Services. In 1992, he received his degree in merchant marine navigation and sea transportation and sailed from 1989 to 2000 in South America, the Red Sea, and between Africa and southern Spain, mainly on ferries and high-speed crafts. In 1998, he received his Master Mariner qualification. Since 2000, he has been working for the Spanish Maritime Safety Agency—agency for Safety and Security at Sea and Maritime Search and Rescue service.

In 2011, he obtained a degree in both clinical, and work and organizational psychology. In 2013, he completed a master's degree in investigative psychology and in 2017, a master's program on statistics for health sciences. In March 2018, he received a PhD in health sciences from Cadiz University.

Main areas of interest:

Analytical research on human factors investigation, featuring studies in the civil marine and Vessel Traffic Services environment. Background includes participation in workshops, conferences, and seminars on maritime safety and human factors. Experience in designing and running studies, and writing scientific reports and research strategies.

Performed, designed, and validated a screening test for Vessel Traffic Services operators based on a situational awareness assessment. Published in the *Journal of Work and Organizational Psychology*, 30 (2014).

Involved in the development of a framework for researching and training in human factors in seafaring, and participant in the International Association of Marine Aids to Navigation and Lighthouse Authorities (an international organization in charge of Vessel Traffic Services) Workshop for Human Factors in Vessel Traffic Services, in Gotenburg, Sweden, in 2015.

Speaker at the Atlantic Stakeholder Platform Conference, Brest, 2015, presenting a joint project with a Canadian team under the title: Human Factor: The Key Element of Maritime Accidents.

Author of "Human Factors in Seafaring: The Role of Situation Awareness," *Safety Science* 93, January 2017.

Reviewer at the *Global Journal of Health Sciences* (Canada).

1 Introduction

For years, technological advancements have fueled a real revolution in the field of navigation, far surpassing any progress on training and the preparation of crews, especially officers. Recently, at the second Atlantic Stakeholder Platform Conference, held in Brest (France), Cordon, Mestre, and Walliser (2015) presented an updated study on the merchant fleets of Euro-Atlantic countries and Canada, analyzing accidents and maritime incidents from 2012 to October 2015. They concluded that 85% were either directly caused or aggravated by human error.

Fortunately, international agencies are becoming increasingly aware of the huge importance of the human factor in shipping accidents. Additionally, shipping companies are recognizing that the training and preparation of officers do not represent a cost, but an investment that increases the security and productivity of ships.

In this regard, in 2010, the International Maritime Organization (IMO) clearly stated that:

> The key to maintain a safe transport environment and keep our oceans clean is all sailors around the world observing high standards of competence and professionalism in tasks they perform on board. The International Convention on standards of training, certification, and Watchkeeping for seafarers 1978, as amended in 1995 and again in 2010, sets these rules, and regulates the granting of certificates and control mechanisms of the guard. Its provisions apply not only to seafarers, but also to owners, training institutions, and the national maritime administrations. (IMO, 2010, p. 9)

However, the greatest stimulus to developing this manual was certainly the latest revisions to the International Convention on Standards of Training, Certification and Watchkeeping for Seafarers (STCW), which included, for the first time, aspects of non-technical training, such as personal, group, and leadership skills. These amendments, known as the "Manila Amendments," refer to

- Bridge resource management
- Management of machine room resources
- Leadership and management skills

Introducing these amendments requires IMO member states to pursue policies of seafarers' training focused not only on the technical aspects of managing the bridge, but also on other responsibilities performed on board.

The Manila Amendments, whose implementation is compulsory for all signatory countries, are being applied unevenly across countries. Some nations have specific preparation courses, *master courses*, while other countries include the subject as part of a regular training course. The fact is that no manual has been compiled that includes everything necessary to prepare these materials with the necessary guarantees, the norm being the dispersion of knowledge with no connection to legislation.

This manual attempts to be a text written by and for mariners, but with the rigor of a psychology text. This has been the primary motivation for this work: the creation of a complete manual, with the thoroughness of a scientific psychology text, but with seafarers in mind, and specifically tailored to their needs. This book is not intended for psychologists (apart from those interested in work and organizational psychology), although it has been written by one. It is possible that for a psychologist some topics lack sufficient depth, but that is not its purpose. The goal is the merchant navy officer's needs, in areas in which modern psychology can help them.

This manual has been authored by a sailor with seamen in mind.

Chapter 2, probably the most profound in psychological concepts, is devoted to leadership. This is because the Manila Amendments place leadership first and it is certainly one of the most important concepts that an officer must focus on. Lately, the study of leadership has become very fashionable, as the proliferation of the phenomenon of coaching and gurus has extended to all types of businesses, as if it were a panacea for almost everything in corporate management.

This has meant a veritable explosion of literature on leadership, especially since the modern approach is centered, above all, on transformational leadership.

However, as the reader will observe, this concept is not the most important.Instead, more attention is paid to models that might be considered more classic, or at least, not fashionable at present. This is because aboard a ship the leadership has restrictions that are not present in a multinational company. For instance, the teams are fixed and normally we will not train them at our will, nor will we make decisions about the governance imposed by companies, nor the embarkation periods, or many other factors that would be necessary if we want to apply the concepts of transformational leadership on board. That is why I believe that contingency models, where leadership adapts to circumstances, are the most useful for seafarers.

Chapter 2 also includes a section on leadership in emergencies, focused on passenger ships, where it is essential to understand the mechanisms of fear and how crowds behave.

Chapter 3 is devoted to teamwork and the study of groups and their phenomena, since it is difficult (as they do in coaching, for example) to explain how teams work without first understanding the dynamics that drive groups, and how people act when they are in groups. A discussion follows on how to build better teams and how to make them work, always with the ultimate goal of job satisfaction and productivity.

Although it is not explicitly mentioned in the Manila Amendments, I consider knowledge of multiculturalism essential to becoming a leader and being able to manage teams, because crews are increasingly more diverse and it is essential to understand how people from different cultures, some radically different, work together. I have also included a section on women and their role on board, since we are seeing an increasing number of women joining the fleet.

Chapter 4 addresses what I consider the most practical part of the manual, the study of the human factor on board ships, which would not be complete without an in-depth examination of the psychosocial factors that a sailor must face. If these are not identified, the rest of this book does not make sense, because it is on these conditions that life and work on board hinge.

To conclude, Chapter 5 looks at the Vessel Traffic Services (VTS), since many sailors become traffic operators. The VTS have some shared characteristics with the workplace on board, but other characteristics are logically different.

All recommendations in this manual are addressed to officers, shipping companies, and national administrations, from a psychological perspective.

I hope that maritime students find this manual useful in completing their studies in this field; included are some tips and suggestions that they can use in their daily lives on board ship. I also hope that teachers in the official training institutions will find all pertinent information in this book rather than having to search through multiple documents, and that it helps them in their teaching tasks.

REFERENCES

Cordon, Jose R, Jose M Mestre, and Jorge Walliser. 2015. "Human Factor: The Key Element of Maritime Accidents (PDF Download Available)". In Atlantic Stakeholder Platform Conference, Brest, France.

IMO. 2010. "STCW Guide for Seafarers."

2 Leadership

Leadership is one of the most recurring topics in human resources literature today. For sailors, leadership is something that occurs naturally on board ship and it is probably the key concept in this type of organization.

If there is any workplace where the concept of leadership has been better developed it is on board a merchant ship, and in the military, where the hierarchy of personnel is even more pronounced. The characteristics of a seafarer's work environment are so special that the concept must be adapted to each work place. The isolation, the interdependence, and the limited availability of resources make a ship a closed system where the concepts of leader and typical leadership that are handled in the conventional literature must be adapted to their own field.

2.1 LEADERSHIP ON BOARD

Leadership development is an increasingly common concept in countries such as the United States and the United Kingdom, in organizations, and in universities where many programs have been devised to develop effective skills for leaders of groups and projects.

Leadership development teaches potential leaders how and when to lead. It teaches them when to be a leader and when to be a follower, how to lead by example, and the dynamics of being a leader within a group.

Leadership development is for people who wear many hats. In the maritime field, leadership has always been a skill that must be learned on board, but with mixed results, since on many occasions there has been a lack of a good role model or simply insufficient inborn leadership skills.

Traditionally in the merchant navy, leadership is imposed by the shipowner who decides a captain's and officers' positions; however, this does not guarantee that the crew's functioning as a team will be effective, especially in emergency situations; we need only remember cases like the captain of the *Costa Concordia*, and his laissez-faire attitude that led his ship to catastrophe and the loss of human lives.

Leadership skills are related to the management and development of people. Among other dimensions, leadership development involves issues of emotional intelligence, situational leadership, and communication and motivational actions.

The Manila Amendments to the International Convention on Standards of Training, Certification and Watchkeeping for Seafarers (STCW) Code (IMO 2010) have finally included what all professionals of the sea have known for centuries, the essential qualities of every officer, such as the skills of teamwork, conflict resolution, leadership, and so forth.

A high-performing leader must be able to manage multiple situations simultaneously. He must be able to motivate his team, creatively build synergies, and keep calm in the face of adversity. In recent US business literature (and to some extent, we can extrapolate it to a ship), a word defines people who possess these qualities, and who intelligently combine diverse personal and professional skills to make things happen, under any circumstances. They are called *horsepower* as a metaphor for the horsepower of an engine.

But in opposition to horsepower, we find those people whose potential is wasted. They are ineffective leaders, refusing the responsibilities of the situation, and are totally undeserving of their position.

Sometimes, it is difficult to determine the ideal profile of a good leader because occasionally they are presented as "superman" who can do anything, are unyielding, and without negative traits. This is not always possible on board ship, because frequently an officer must deal with teams who are difficult to understand (referring to language or culture), are older and more experienced than the officer himself, and perhaps have a greater knowledge of the ship.

A good way to choose a leader's profile is to consider the dynamics of a front-runner. Whereas a passive leader may have difficulty making decisions or he may be hard-headed, a dynamic leader is self-aware, proactive, open to feedback, eager to improve, takes risks, and learns from adversity. As a successful entrepreneur, a dynamic leader sums up the qualities of a good leader.

In his book *Driving People is as Difficult as Driving Cats: Leaders can be Done*, Warren Bennis (2001) describes 10 characteristics of dynamic leaders.

1. Self-knowledge. They know their talents and how to better employ them. Many of them have spent a period of their lives in another country, and that experience has enriched them as individuals. They are independent and question their own assumptions and beliefs. Various life experiences have taught them to think about who they are and what they are.
2. Openness to feedback. This is a skill that they respect and value. They have an internal valuable source of reflection and constructive criticism that forces them to progress and improve. This same personal trait is looked for and valued at work.
3. Eager to learn and improve. They are great inquirers and interlocutors, and always seek to improve. They are open to change, receptive to new information, dislike being surprised by the unexpected, and are eager to acquire new knowledge.
4. Curious and take risks. They are intrepid and bold. For them, it is not only the destination that is important, but also the journey. They are fascinated by new thoughts and ideas.
5. Focused on work. They have great tenacity. They use ingenuity at work. Sometimes, this quality is difficult to uncover during a simple chat, but it can be confirmed by knowing the person in depth. They may not be very good at interpersonal relationships, and may even leave a bad impression, but they are extraordinarily effective in their own field.
6. Learn from adversity. They may have suffered several failures in their lives. Some may have had a difficult childhood or suffered from chronic diseases. The dynamic leaders studied by Bennis faced adversity at an early stage in

their lives. That learning is invariably transferred when they teach others, through anecdotes and advice.

7. Balance tradition and change. They distinguish between and know how to adapt to both a conservative and a changing environment.

8. An open style. They have innovative ideas to teach others, analyze the competition, and ask collaborators for their thoughts.

9. Work well within systems. They work as a team and rely on the processes and structures of the organization to solve their problems.

10. Good mentors and role models. Others use them as examples because they are reliable, and others know that by emulating their steps they will achieve success in whatever they do. They are respected.

2.1.1 Definitions of Leadership

Leadership is a popular concept with multiple definitions, theories, and analyses. Many authors have focused on the profitable jobs in the field, including Solano et al. (2007). It is almost impossible to agree on a single definition for leadership, since different authors approach it from diverse perspectives, mainly as a trait, as a process of influence, or as a behavior.

Yukl, Gordon, and Taber (2002) offer a complete analysis of the concept and identify some key findings; however, they go beyond the objective of this manual, but can be consulted by the curious reader. These authors point out how the concept has evolved from a behavior centered on relationships and tasks to interest in people and production, and claim that this vision has dominated three decades of research. Formerly, the focus has been on leadership in the transformation of organizations. Finally, these authors propose a definition of leadership as a process to influence people and to achieve certain objectives.

Following Solano et al. (2007), these authors identify the characteristics that are common to all definitions, which we can adapt to the maritime field:

- Leadership is a process. The leader's behavior is a continuous flow of interactions between him and his subordinates in the closed space of a ship. Not only must the captain give orders, but he must be willing to listen to his crew, especially to his officers. The captain and the officers occupy formal leadership positions, that is, they are imposed by the shipowner, but this does not mean that they are effective leaders, since this role (leader) is born from daily interactions with their subordinates.

- Leadership as an influence on others is not merely understood as a hierarchical imposition, but as a relationship of trust. Hierarchy on a ship exists because the officers' knowledge and experience are superior to their subordinates. However, this is not always the case, as sometimes, subordinates with great experience may have more knowledge in certain areas than the Officer himself.

- Leadership occurs in a group and involves the achievement of objectives: the fundamental concept for the group is the common objective. Complying with a schedule, with quality standards at work, or with the demands of a charterer are objectives that the captain must transmit to all his crew. They cannot only be the concern of the officers, but the entire crew must be involved as a necessity for their attainment.

2.1.2 FOCUSES CENTERED ON THE LEADER, TRAIT, OR PROCESS?

Since the dawn of research on leadership there has been a discussion about whether it is something innate in a person (trait) or a phenomenon arising from the environment. If we choose the trait approach, we would conclude that leaders are born and thus training is useless. If we choose the second view, everything depends on the leader's context. Experience tells mariners that both approaches are basically true, since there will always be natural leaders; however, many leadership behaviors are learned on board, and only by directing crews.

At this point, it is necessary to differentiate a leader from a mere manager, since the latter does not care about motivating a team but is only interested in relaying the owner's orders. Unfortunately, every sailor has known officers, and even captains, who limit themselves to demanding the fulfillment of a series of objectives without caring about how such objectives are achieved.

Following an exhaustive review of published works, Bass and Stogdill (1990) concluded that it is impossible to distinguish leaders' behavior in any situation; however, this conclusion was later nuanced:

1. There are many features that differentiate those who are leaders from those who are not: physical energy, intelligence greater than the average intelligence of their followers', self-confidence, motivation to achieve, and power.
2. The relevance of the situation influences the leader's behavior and effectiveness.
3. The expression of individuals' behavior determines the influence they have on the leader's behavior. Leader's conduct is mediated by strong norms or styles on rewards and punishments.

As affirmed by Solano et al. (2007), the concept of leadership is related to the concepts of power and authority. Power is the ability to influence people; it is a dynamic concept because it depends on the people over which power is exercised. Authority, conversely, is a right acquired by or granted to some people, enabling them to exercise power. A person may hold authority, for example, the captain who is the representative of the shipowner, but he may be unable to exercise that power because he cannot influence his subordinates. Conversely, natural leadership can also occur in people who have no authority, that is, people who exercise power without holding any position. Research suggests both concepts are related and tend to be confused in leaders.

2.1.3 THE TRAIT APPROACH

This is the most natural and intuitive vision to understand leadership, as something innate in some people; however, the research has been somewhat contradictory, as some authors have claimed that there are underlying traits to all leaders, which are not always visible.

2.1.4 THE LEADERSHIP STYLES APPROACH

What makes this approach interesting is determining the leader's behavior and the consequences it has on his followers. The origins of this approach are found in the works of Lewin (1939), on democratic and autocratic leadership in groups of children. Investigations by many other authors have reinforced and deepened this view. The research concluded that there are two dimensions linked to leadership behavior:

- Structure initiation: Planning work, defining responsibilities, arranging tasks, etc.
- Consideration: Concern for followers' welfare and the maintenance of a good working relationship within the group.

These two dimensions appear independent, that is, a leader can score high in one and low in another, so the circumstances of the situation have a great part to play.

2.1.5 CONTINGENCY MODELS

Contingency is the possibility of something happening that cannot be predicted. An organization must act accordingly to the situation it faces, therefore, there are no universal prescriptions for all scenarios.

According to these models, leadership depends on the situation and the circumstances. Possibly, the theories that arise under contingency models are those that best fit the analysis and leadership that interest us. These models have emerged to overcome the limitations of other approaches where the variables of the situation are not considered. These models have generated significant research and we will deal with them more thoroughly.

The main parameters that we must consider when analyzing the leadership style that best fits any given situation are specified by each organization's internal and external factors.

1. Internal factors:
   - Organization size: In our case, it is extremely variable, since operating a tugboat with six crew members is not the same as operating a cruise ship with 2500 crew members. Obviously, the same leadership style cannot be applied in both structures.
   - Technologies in routine tasks: In the same way, the level of automation and technological standards are highly variable throughout the world fleets, with remarkable differences between different countries and geographical areas.
2. External factors:
   - Environmental uncertainty: Refers to the certainties about the organization itself, that is, a ship owned by a large shipping company with a strong financial backing is in a different position to that of a company from a developing country that may or may not be financially bankrupt.

- Individual differences: Individuals are highly variable in terms of their motivation and expectations, even more so in the current context of multicultural crews who may have different ways of understanding work and life itself.

We will analyze the different models that arise from this theoretical approach and their possible application to the maritime context.

2.1.5.1 Fielder's Model

The key to this model is to identify the best combinations of leadership style and situation, since the effectiveness of an approach directly depends on the situation and all the associated factors where it is intended to apply (Fiedler 2006). In this way, an officer must know how to identify the precise situation he is facing and apply a specific style of leadership. The situation depends on three factors, in order of importance:

1. Relationship between the leader and the subordinate: Degree of trust, loyalty, and respect.
2. Structure of the task: Clarity and level of arrangement; the more formalized it is, the greater the officer's power of command.
3. Position of power: Authority inherent in a leader; in our case there is a strong relationship between this and the authority imposed by the rules and legislation of each country. However, the rules remain constant for everyone and emanate from the captain as representative of the owner, down in a pyramidal structure.

Fiedler used a sociogram to illustrate preferred work partners. He found that the best leaders were those focused on human relationships, while the least preferred leaders focused on tasks. The model is structured as it appears in Table 2.1.

According to this model, there is no single successful leadership style, but it depends on the task and the circumstances; therefore, two styles of leadership are defined: the task of motivation and the motivated relationship. The leader must know how to adapt his style to the task and the group, for example, an impersonal leader who is task oriented can be effective when handling a large crew with structured and clearly defined tasks, such as modern cruise ships. However, this same leader must adapt his style when dealing with his group of officers, since this group is small and the tasks are not as rigidly structured or defined. Table 2.2 details this issue.

2.1.5.2 Hersey and Blanchard Situational Leadership Model

Also called the theory of situational leadership, this model focuses on analyzing the disposition of followers, reflecting the leader's acceptance or rejection in the organization; therefore, the effectiveness of the leadership depends on the behavior of his followers (Hersey, Blanchard, and Natemeyer 1979). In this way, the leader must adapt his style to the conditions of his subordinates. It is a simple model that serves to quickly adapt the leadership style to a new situation.

Four leadership types are defined, according to the leadership orientation toward the task or toward the relationship, as reflected in Table 2.3.

TABLE 2.1
Fielder's Model

Category	Leader–Member Relationship	Structure of Tasks	Power	Favorable Situation
I	Good	High	Strong	Very favorable
II	Good	High	Weak	Very favorable
III	Good	Low	Strong	Very favorable
IV	Good	Low	Weak	Moderate
V	Deficient	High	Strong	Moderate
VI	Deficient	High	Weak	Moderate
VII	Deficient	Low	Strong	Very favorable
VIII	Deficient	Low	Weak	Very favorable

The application of the model is developed in six steps:

1. Identify the current task so that it can be developed as efficiently as possible.
2. Find the skills and knowledge necessary to carry out the task.
3. Determine the level of knowledge and competence of each team member.
4. Fix the level of motivation of each person.
5. Evaluate the level of adaptation of each member to each position.
6. Apply the right leadership style.

In the case of a ship, this model can be difficult to apply, since roles and positions are usually determined in advance, and the structure of the team of subordinates is already determined by the company or by practice, and it can be difficult to modify these structures. But sometimes it is possible to adjust the team following this system.

The final stage of the model, once the first six steps have been evaluated, is to determine the "disposition of the follower" as one of four types:

D1: Not competent and unreliable or does not wish to assume responsibilities.
D2: Motivated, but not very competent and ill-prepared.
D3: Ready, but without desire to do what is asked.
D4: Ready and willing to follow the leader.

The model also includes a typology of leadership, and the leader must decide which one best fits:
E1: The leader orders:

- Actively participates in tasks, although with little personal involvement.
- Controls exhaustively because personnel is not reliable.
- Gives precise orders.
- Supervises exhaustively.
- Makes all decisions.

TABLE 2.2
Leadership Styles Depending on Context: Fielder's Model

Situation	1	2	3	4	5	6	7	8
Leader/followers relationship	Fine	Fine	Fine	Fine	Poor	Poor	Poor	Poor
Task repetitiveness	Repetitive	Repetitive	Non-repetitive	Non-repetitive	Repetitive	Repetitive	Non-repetitive	Non-repetitive
Power	Strong	Weak	Strong	Weak	Strong	Weak	Strong	Weak
Leadership style	Task oriented	Task oriented	Task oriented	Oriented to relationships	Oriented to relationships	Oriented to relationships	Oriented to relationships	Task oriented

TABLE 2.3
Leadership Styles as in Situational Leadership Model

Leadership Styles	Communicate	Sell	Participate	Delegate
Leader's behavior	Decide, define tasks, communicate where and when to perform them	Define tasks and convince subordinates	Decisions are shared	Leader not necessary
Task oriented	High	High	Low	Low
Oriented to relationships	Low	High	High	Low

E2: The leader persuades:

• Directs and supports.
• Explains his decisions.
• Accepts clarifications.
• Teaches and motivates crew.
• Convinces through his actions and his vision of things.

E3: The leader participates:

• The crew is very prepared, so it is necessary to involve them.
• Active communicator.
• Motivates crew.
• Shares decisions and motivations.
• Establishes public favor–debt relationships with his followers.

E4: The leader delegates:

• Directs and supports, but from a distance.
• Observes and supervises.
• Promotes autonomy.
• Values experience and skill.
• Shows confidence.

2.1.5.3 Model of Leader Participation

This is a normative model, since it provides a series of rules that the leader could follow in different situations (Vroom and Yetton 1973). This model proposes five situations and a decision tree to determine which one to use in every case:

1. Autocratic Process I (A1): The leader makes the decision using the available information.
2. Autocratic Process II (A2): Information is requested from the team, but they do not participate in the decision-making.

3. Consultative Process I (C1): The decisions are explained to each member of the group individually, not in groups, but the leader decides.
4. Consultative Process II (C2): The decisions are explained in groups, but they are taken by the leader.
5. Group II Process (G2): The group discusses and decides, the leader acts as facilitator.

The method to determine which one to use is to answer questions and choose the path according to the answers. The questions are as follows:

1. Is the technical quality of the decision important?
2. Do you have enough information to decide for yourself?
3. Is the problem defined and structured to reach a known resolution?
4. Does the team have to accept the solution to make it work?
5. If you decide alone, will the team accept your decision?
6. Is the team motivated by your goals?
7. Is it possible that there is disagreement in the team to reach a decision?

The decision mechanism is shown in Table 2.4.

2.1.5.4 Path to the Goal Model

This model is inspired by the expectation theory of motivation, which states that the leader's role is to help his followers achieve their own goals, as long as the objective is compatible with that of the organization (Shamir, House, and Arthur 1993).

Four types of leadership are identified:

1. Managerial or directive leader: The followers know what is expected of them, but the leader programs and directs the activities.
2. Supportive leader: The leader is regarded as a friend, showing interest in his followers.
3. Participative leader: The leader consults and shares his decisions with his subordinates.
4. Achievement-oriented leader: The leader seeks the highest performance from his subordinates, proposing challenging objectives.

The model emphasizes subordinates' goal achievements through accomplishing personal objectives. The leader helps clarify and attain these goals, thereby increasing his subordinates' job satisfaction. The leader must clarify any ambiguous goals and focus the expectations of his subordinates. In this sense, the term *trajectory goal* is used; the leader's behavior will prove motivational if his subordinates gain satisfaction from their performance and provides them with direction, training, support, and rewards on successfully completing the objectives.

The hypotheses derived from this model are

1. Leadership yields greater satisfaction with ambiguous tasks.
2. If the tasks are structured, the best style of leadership is support, as it provides greater satisfaction.

TABLE 2.4

Decision Process: Leader Participation Model

1: Is the Technical Quality of the Decision Important?	2: Do You have Enough Information to Decide for Yourself?	3: Is the Problem Defined and Structured to Reach a Known Resolution?	4: Does the Team have to Accept the Solution to Make it Work?	5: If You Decide Alone, will the Team Accept Your Decision?	6: Is the Team Motivated by Your Goals?	7: Is it Possible that there is Disagreement in the Team to Reach a Decision?
No	—	—	Yes	No. The leader makes the decision using the available information.	—	—
—	—	—	Yes	Yes. The group discusses and decides, the leader acts as facilitator.	—	—
Yes	Yes	—	No. The leader makes the decision using the available information.	—	—	—
—	—	—	Yes	Yes. The leader makes the decision using the available information.	—	—
—	—	—	—	No	Yes. The group discusses and decides, the leader acts as facilitator.	—
—	—	—	—	—	No	Yes. Decisions are explained in groups, but they are taken by the leader.

(*Continued*)

TABLE 2.4 (CONTINUED)
Decision Process: Leader Participation Model

1: Is the Technical Quality of the Decision Important?	2: Do You have Enough Information to Decide for Yourself?	3: Is the Problem Defined and Structured to Reach a Known Resolution?	4: Does the Team have to Accept the Solution to Make it Work?	5: If You Decide Alone, will the Team Accept Your Decision?	6: Is the Team Motivated by Your Goals?	7: Is it Possible that there is Disagreement in the Team to Reach a Decision?
—	—	—	—	—	—	No. Decisions are explained to each member of the group individually, not in groups; but the leader decides.
—	No	Yes	Yes	No	No. The group discusses and decides, the leader acts as facilitator.	Yes. Decisions are explained in groups, but they are taken by the leader.
—	—	—	No. Information is requested from the team, but they do not participate in decision-making.	—	—	No. Decisions are explained to each member of the group individually, not in groups; but the leader decides.
—	—	No	No	No. Decisions are explained in groups, but they are taken by the leader.	—	—
—	—	—	Yes	Yes. Decisions are explained in groups, but they are taken by the leader.	—	—

(Continued)

TABLE 2.4 (CONTINUED)
Decision Process: Leader Participation Model

1: Is the Technical Quality of the Decision Important?	2: Do You have Enough Information to Decide for Yourself?	3: Is the Problem Defined and Structured to Reach a Known Resolution?	4: Does the Team have to Accept the Solution to Make it Work?	5: If You Decide Alone, will the Team Accept Your Decision?	6: Is the Team Motivated by Your Goals?	7: Is it Possible that there is Disagreement in the Team to Reach a Decision?
—	—	—	—	No	Yes. Decisions are explained in groups, but they are taken by the leader.	—
—	—	—	—	—	No	Yes. The group discusses and decides, the leader acts as facilitator.
—	—	—	—	—	—	No. Decisions are explained in groups, but they are taken by the leader.

3. Executive leadership will be perceived as redundant in the case of highly experienced subordinates.
4. The clearer and more bureaucratic the relationship of authority is, the better it is to apply supportive leadership.
5. Directive leadership is most effective when there is conflict within the group.
6. Followers with an internal locus of control feel more comfortable with a participatory leadership style.
7. Subalterns with an external locus of control prefer a managerial style.

The locus of control is a personality trait that represents the attribution that a person makes about the results he obtains depends on his efforts. There are two types: internal control locus and external control locus.

The internal control locus occurs when an individual perceives that an event directly depends on his own behavior. That is, the person believes that what has occurred is thanks to his behavior and that he has control over the external consequences.

The external control locus occurs when the individual perceives that an external event took place independent of his behavior. Therefore, the individual associates the event that occurred with chance, luck, or destiny.

Table 2.5 shows House's route–goal model.

2.1.6   LEADERSHIP MODELS FOCUSED ON FOLLOWERS: IMPLICIT LEADERSHIP THEORIES

These models are based on the idea that both leaders and followers have a script or stereotype about the expected behaviors of a person who is to be considered a leader (Wofford, Goodwin, and Whittington 1998). These theories define the beliefs

TABLE 2.5
Route–Goal Model

Situation	Followers Lack Self-Confidence	Lack of Interest in Work	Setting Ambitious but Feasible Goals	Ambiguous Work
Leadership style	Supporting	Goal oriented	Participative	Directive
Impact on followers	Increased security in tasks	Setting ambitious but feasible goals	Followers suggest and participate	Clarity on how to obtain rewards
Results	Greater effort, less conflicts, better performance and satisfaction	Higher performance and satisfaction	Less turnover, better performance and satisfaction	Better performance and satisfaction

Source: Adapted from Hellriegel and Slocum (1998).

about what a leader's behavior should be to be considered for such a role, and what is expected of him.

It has been empirically demonstrated that people use an attribute assignment process to identify leaders with the ideal qualities of a leader (Rosch 1999). In this way, the concept of leadership falls more on the followers than on the leader himself, that is, a person will be perceived as a leader if the group perceives in him the characteristics of the role model.

In this model, leadership is organized into three levels or categories:

1. General or supraordinal level: Only distinguishes between leaders and non-leaders.
2. Basic level: Leaders are classified according to the environment, e.g., political, military, or religious.
3. Subordinate level: Basic levels are subcategorized, such as in the military, e.g., captain or sergeant.

The differences between the implicit ideas that subordinates have about leadership and the leader's own behaviors determine the status of the leader. In addition, these theories or implicit ideas model followers' perception of a leader. Several studies support these theories and show how different populations have different perceptions of leadership (Castro Solano 2006). For example, the military looks for more heroic behavior, but for civilians the dimensions of ability and experience are predominant.

The organizational culture also influences the perception of a leader. In transactional cultures, the directive styles prevail, oriented to the norms, while in organizations with a transformational culture the leader must be more participatory and democratic.

If the perception of the leader's behavior coincides with that expected by his followers, he will be accepted; otherwise, he may be rejected, and although he may impose authority, his actions could be ineffective.

2.1.7   LEADERSHIP AS A PROCESS OF INFLUENCE: IDIOSYNCRATIC CREDIT THEORY

For Hollander (1985), leadership is a process of influence between two or more interdependent people to achieve the goals of a group. This relationship is built over time and involves an exchange, with the leader providing resources to the group, thereby giving him status and influence. This theory is based on the fact that leadership is not static and it is a two-way process.

This theory attempts to explain a leader's deviations from the norm and his legitimacy in influencing the group, even if it leads them to defy established norms. The expression *idiosyncratic credit* refers to using the leader's status given by the group to implement innovative behaviors. In this way, these deviations from the norm will be better tolerated in later attempts to influence the group. These processes can be clearly observed in the political context, where leaders who are perceived competent can deviate from their electoral promises and this behavior will be accepted by their followers.

The last important concept is legitimacy. Hollander proposes four origins of legitimacy:

- Initial conformity to the rules: The leader's legitimacy will be greater if he adapts quickly to the norms of the group.
- Origin of authority: External imposition or election by the group, with election providing greater legitimacy.
- Competition: The greater the competition, the more his legitimacy.
- Followers' identification with the leader: If the leader shares the same ideals and values as his followers, he will obtain greater legitimacy.

2.1.8   CHARISMATIC LEADERSHIP

Among the new theories on leadership, the charismatic and the transformational theories stand out (although the latter can be considered a derivative of the charismatic). We have already spoken briefly about transformational theory and we will deal it with in depth later on (see Section 2.1.9).

Charisma is a controversial concept in psychology because it has esoteric connotations and is difficult to define. However, it is a construct that is easily distinguishable in some people who are naturally gifted.

The characteristics of charismatic leaders are

- Love life: They are cheerful and optimistic people. They do not lament about unimportant things. They are passionate about what they do.
- Value people's potential: They expect the best from their subordinates; they see people not as they are, but who they could become.
- Generate hope: They set challenging but realistic goals for their group.
- Transmit hope to people: They convey conviction about the future.
- Offer and generate sacrifice: They are role models and convince others that their sacrifice will be rewarded.
- Offer themselves to others: They are generous, sharing their knowledge, resources, and time with their followers.

How does one develop charisma? As specified, charisma is a trait that some people are born with, while others can enhance it by knowing some key points:

- Observe others, especially people with natural charisma, and try to imitate certain attitudes, including their way of listening, their look, or their smile.
- Be open to and proactive in interpersonal relationships. Try to take the lead in conversations, proposing topics of conversation and conversing on your own terms, but always showing interest in the opinions of others and giving them time to express their ideas.
- Smile, and keep eye contact with your interlocutor, which generates confidence.
- Charisma is mysterious, it keeps some of your privacy, a halo of reserve over your personality that makes it more attractive.

- Transmit good ideas to the group, propose interesting things to do, places to go, etc.
- Choose a personal style, something unique that defines you and clearly distinguishes you from others; it can be a personal image or some salient (and positive) personality trait.

2.1.9 TRANSFORMATIONAL LEADERSHIP AND TRANSACTIONAL LEADERSHIP

It was Burns (1978) who proposed this topology about leadership, distinguishing between a transactional and a transformational style based on leaders and followers exchanges.

Transformational leaders provide a purpose that transcends short-term goals and focuses on higher-order internal needs. Comparatively, transactional leaders focus on the exchange of resources. Namely, the transformational leader seeks what his followers have identified as their needs, the transaction providing access to an exchange of resources and offering something in exchange for followers to do his will.

However, it was Bass (1985) who developed the theory of transformational leadership, commonly accepted and more elaborated, mainly by his idea that these are independent concepts and not the extremes of a continuum, as proposed by Burns. For example, a leader can have both characteristics but to different degrees, and a good leader must combine both styles. In addition, Bass developed a theoretical system with eight dimensions, including three styles:

- *Transformational leadership* includes four factors:
  1. *Charisma or idealized influence* is a leader's trait that emphasizes trust, does not hide from difficult decisions or complicated issues, is committed to the task, is aware of the scope of his decisions, and establishes relationships with his followers at an emotional level.
  2. *Inspiring motivation* is when a leader implants a vision of the future in his followers, he challenges them to improve their performance to the maximum; they speak with enthusiasm, optimism, and motivation, giving meaning to his decisions.
  3. *Intellectual stimulation* is the desire for risk-taking and the motivation to search for original solutions, encouraging their followers to express their ideas, thereby stimulating creativity.
  4. *Individualized consideration* is the quality of a leader who treats his followers as individuals and takes their abilities and aspirations into consideration. He knows how to listen, advise, and teach.
- *Transactional style* includes three factors:
  1. *Contingent reward* is when relations are an exchange of fair, constructive, and positive rewards, the leader recognizing the effort of his followers.
  2. *Administration by exception* refers to the degree to which the leader undertakes corrective actions with his followers. Some authors (Hater and Bass 1988) discriminate between active and passive management. Active leadership tends to anticipate problems by monitoring

subordinates' actions; however, the liabilities are reactive, thus the leader waits for the problem to arise before making decisions.

3. Finally, the *laissez-faire* style, not leadership or non-intervention policy, which results in the absence or avoidance of leadership. Consequently, it is a style in which supposed leaders abdicate their responsibilities, are reluctant to make decisions, doubt the time to act, and are absent when needed.

2.1.9.1 Transformational Leadership

Research by Bass (1985) and Bass and Stogdill (1990) states that there are three types of leadership: the transactional (exchanges between the leader and his followers), the transformational, and the *laissez-faire*. Usually, the transactional produces positive effects on performance and job satisfaction; however, the transformational also influences the beliefs and attitudes of subordinates, causing deeper changes.

The transformational leader can make subordinates transcend their own interest for the benefit of the group and alter their own hierarchy of needs. He can achieve important changes in the ideals, beliefs, and attitudes of his followers and increase their performance. Bass has studied his model through the multifactor leadership questionnaire (MLQ), which allows the two types of leadership to be measured. These traits are independent and manifest themselves in different ways in each leader.

Transformational leadership factors:

1. *Charisma*. It's the most important factor. The leader must get his followers to identify with him, especially with his goals and objectives.
2. *Inspiration*. The ability of the leader to motivate or excite his followers about the importance of achieving the proposed objective or mission.
3. *Intellectual stimulation*. The leader must motivate his subordinates to use their own resources to solve problems autonomously.
4. *Individualized consideration*. The leader should treat each person individually, knowing their resources and anticipating their difficulties and limitations.

The research appears to consistently support Bass's hypothesis: transformational leaders achieve higher performance from and greater satisfaction in their subordinates than leaders who use only the exchange of rewards (transactional).

Leaders who obtain the worst results are those who do not lead—those who follow a *laissez-faire* style.

The theory explains the process of producing profound changes in followers, more than a mere exchange of rewards. While this model is probably the most in trend, it is not exempt from criticism.

2.1.10 Leadership and Intergroup Relations

Groups are never isolated but maintain relationships with similar or different groups. Case in point are crews, where these processes are mediated by the isolation of the ship and the rotation processes for vacations, breaks, transfers, etc. However, it is important to know how group dynamics influence leadership. Sherif (2010) is probably the most influential author in the field, with his studies on groups of children in a camp, observing how competitive activities increased group cohesion and how individuals with better dispositions emerged as natural leaders. Various studies have shown these processes are valid in all groups.

The current perspectives on intergroup interaction affirm that leadership must be studied from the perspective of the group itself and its relationships with others. When categorized as members of a group, people as members of that group tend to perceive themselves as similar and different from members of another group.

This phenomenon can be observed in politics and sports, but there are studies where mere random assignment establishes these relationships. In this way, the group leader will be the role model, who can best identify with the values of the group.

The theory of social identity advocates that the more that people are identified with the larger group, the greater the adjustment between role model and leader will be and the greater the perceived group effectiveness will be. The leader model is not something static, but it will vary depending on the group's objectives. An effective leader must adapt his leadership style to the role model that best defines the group in each situation.

2.1.11 Intergroup Phenomena

It is important to know the characteristics of intergroup relations, because groups are formed naturally and leadership processes emerge in the same way. If we are managing large or intercultural crews, these processes are necessary and it is important to know how to identify them to deal with crews in an appropriate manner. Knowledge of the intergroup hostility process is of great importance in understanding the relationships between groups of people. There are three types: stereotypes, prejudices, and discrimination, which sometimes get confused.

1. *Stereotypes* are commonly defined as consensual beliefs about the attributes (personality characteristics, behaviors, values) of a social group and its members. Thus, they are perceptions about a person simply because of his membership of a group. It is a widely debated concept, with some authors considering it strict and erroneous, while others think it contains some truth and that the perceptions can arise from normal cognitive processes, not pathological ones. What is clear in the literature is that stereotypes about outgroups and minorities tend to have more negative connotations than stereotypes about groups themselves and the majority (Hilton and Von Hippel 1996).

2. *Prejudices* are derogatory attitudes toward a person due to their membership of a social group. Here, the consensus is that prejudice is a negative intergroup attitude. One of the most influential authors in this field defines prejudice as "having social attitudes or derogatory cognitive beliefs, expressing negative affect or presenting discriminatory or hostile behavior towards members of a group due to their belonging to that particular

group" (Lepore and Brown 1997). Prejudice, therefore, is based on beliefs about the attributes of the social group that are likely to provoke social rejection. In recent years, it has been reported that intergroup hostility expressions have been decreasing in Western countries; however, in reality, they are transforming into or hiding under other forms of hostility, including symbolic or ambivalent racism. Other authors defend the distinction between open prejudice and subtle prejudice, the former being based on ethnic differences and the latter more on cultural differences (Pettigrew and Meertens 1995).

3. *Discrimination* is understood as relationships that produce differentiation in access to limited resources. It refers to norms and practices that limit this access to certain people due to their belonging to a determined social group. In a more individual context, it is the different treatment that a person receives because of being different. This phenomenon has been widely studied and its existence has been empirically demonstrated.

2.1.12 Intergroup Hostility

It is important to know the mechanisms that lead to hostility between groups, especially in the case of large crews and/or multiculturalism on board ship. We can distinguish between individual and group or contextual factors.

As individual factors:

- The personality characteristics of the individuals, that is, their personality.
- Their feelings of frustration and deprivation, when an individual perceives they are being treated unfairly, tending to explain it by external factors, including discrimination, rather than internal factors (insufficient proficiency, for example).
- Social comparison processes, motivated by the psychological need for justice, control, knowledge, self-affirmation, and belonging. These are natural feelings that define us as social animals.

As cognitive processes:

- Categorization and self-categorization, which are the processes by which an individual is placed within a group; they can be directed externally, but also internally, as they are feelings of self-belonging.
- The tendency to homogenize the outgroup is the standardized outer view of the other groups, tending to think that because the groups share certain characteristics, they must share others; in addition, these are usually negative characteristics.
- The selective perception of stimuli and their relationships (*illusory correlation*), that is, the tendency to process information from the outgroup that confirms our beliefs about it, discarding those that may contradict stereotypes, and inclining to attribute events that occur to the outgroup as a result of these characteristics, instead of the circumstances or the prejudices. For

example, if the opposing team loses a competition, the result tends to be attributed to their insufficient training, when maybe arbitration decisions caused their defeat.

Attributional biases are closely related to the previous concepts. When we speak of attributional biases, we are referring to the tendency to attribute behavior exclusively to the dispositions (internal, for instance, expertise, knowledge, will, etc.) of the person and to ignore the power of the situation determinants and overestimate the behavior consistency of individuals through different situations (Nisbet and Ross 1980).

There are authors who claim that people have a certain tendency to make an attribution that favors themselves so that all the factors in a situation are not considered (*situational bias*). Many of these biases are socially learned.

One of the most important biases is the *fundamental attribution error* bias; there is a tendency to attribute other people's behavior to internal causes than to external situational reasons. By means of this effect, the observer-actor tends to attribute his own outcomes to situational factors, while if the same behavior is observed in another person, there is a tendency to attribute the behavior to internal causes (dispositional attributions). In the biases of self-benefit, people usually make internal attributions for their successes and external for their failures, to maintain self-esteem.

It is useful to know all these mechanisms, both individual and group, if we want to understand people's behavior in groups and anticipate problems arising from the interaction between groups.

2.1.13 INTERGROUP HARMONY

In the same way that we are interested in the mechanisms that produce hostility between groups, it is important to know how to promote good relations between them.

Allport's (1954) theory of intergroup contact is the most important in this field. It postulates that contact between groups can reduce hostility, although certain conditions are necessary, such as situational factors, frequency, quality, variety, and atmosphere. Allport points out four factors or conditions for optimal intergroup contact:

- Equal status of the participants in the interaction.
- The achievement of common objectives.
- Cooperation.
- Institutional support (in standards, sanctions, and regulations).

There is abundant empirical evidence on this theory's validity and the importance of the four factors has been proven in research on several continents. According to these investigations, it seems that processes of categorization or recategorization take place. Once outgroup members have been contacted, they are recategorized as they are now known more as individuals than as group members.

This intergroup contact will not always be easy to carry out in the context that concerns us; however, the common objectives, cooperation and the creation and implementation of internal regulations that avoid intergroup hostility, can always be achieved.

Evacuation training exercises for passenger ships, for example, can be a good time to implement these processes. Similarly, consideration should be given to the implementation of time sharing where intergroup cooperation is reinforced and common goals are achieved, such as passenger satisfaction or compliance with schedules and the shipping company's quality standards.

2.1.14   LEADERSHIP AND ORGANIZATIONAL CLIMATE

Leadership is a phenomenon that has the greatest influence on the climate in organizations. Several studies have determined that a good work environment among coworkers is a compromise for family life, social benefits, job satisfaction, and leadership. Leadership influences not only the organizational climate, but also productivity.

Transformational leadership advocates a small distance of power (i.e., closeness in relationships with subordinates), individual attention to the needs of each member, and the offer of motivational goals and rewards, thereby stimulating the organizational climate. Research on this context indicates that the transformational leadership style has the greatest impact on the climate in organizations (Berson and Linton 2005). It is important, therefore, to foster relationships based on trust, closeness, and honesty.

2.1.15   LEADERSHIP AND JOB SATISFACTION

In recent years, job satisfaction is one of the most studied variables, given its importance in terms of the impact it has on research. As a consequence of job satisfaction, subjective well-being, commitment, and positive behavior toward the company can be cited (Edwards et al. 2008). Background factors have been identified (among others): a motivating job, perform tasks under favorable conditions, a fair system of rewards, and perform a job compatible with the person's personality (Omar 2011).

Confidence in an organization is based on trust in the immediate superiors. In this way, a transformational leadership style is indicated to generate trust in the employees. Omar (2011) found empirical evidence that when employees perceive a person as a transforming leader, he increases their job satisfaction, especially the facets of idealized influence (charisma, vision, and decision) and intellectual stimulation (making decisions through reason). They also found a relationship between trust in superiors and transformational leadership and between trust and job satisfaction. Other studies also indicate that there is a significant association between job satisfaction and satisfaction with life in general (Paris and Omar 2008).

There is no doubt that an increase in job satisfaction represents a competitive advantage for companies and should be encouraged by the organization itself, although many factors implicit in this concept are beyond our control (including the compensation of crew members). Others styles, such as the leadership style, we can control; in this sense, the adoption of a transformational leadership style is the most appropriate for this purpose.

2.1.16 Cross-Cultural Leadership

In the context of merchant shipping, multicultural groups are employed on a daily basis and it is easy to find crews composed of many nationalities with different languages and characteristics. These variances will affect the behavior of the leader and will force him to adapt his leadership style, so a directive style could be counterproductive for groups of Western crew members, but not for Oriental people. Hofstede (1983) found four dimensions that serve to characterize different cultures and to establish comparisons between them:

1. *Distance of power*. Acceptance of an unequal distribution of power in an organization. It is assumed that an authoritarian leader in a society with a low-power distance would prompt his rejection by his subordinates. Whereas in a society with a high-power distance, a non-directive leader would be perceived as weak and ineffective. These distances are observed between Western and Eastern cultures (see Section 3.6.2).
2. *Individualism/collectivism*. Individualism is typical of societies in which ties between people are weak, as in Western and especially Nordic and Anglo-Saxon societies. Collectivism is typical of cultures in which people are integrated into strong and cohesive groups that protect them in exchange for unquestioning loyalty, typical of Oriental nations. In a collectivist culture, a leader focused on the task will have some difficulties. On the other hand, in an individualistic society, a relationship-centered leader will face different problems, as individuals tend not to establish intimate links easily.
3. *Masculinity/femininity*. In male-centered societies, gender roles are clearly defined. Men are expected to be assertive, tough, and focused on material success. Women are expected to be modest, tender, and interested in the quality of life. In female societies, both roles overlap to a large extent. Likewise, there are great differences between Western and Eastern cultures in this aspect, and an officer must know how to handle such differences. Hofstede states that male and female cultures create different types of ideal leaders. In male societies, the leader must be assertive, determined, aggressive, and make decisions for himself, without consulting the group. In female societies, the ideal leader is more intuitive than rational and seeks consensus.
4. *Avoidance of uncertainty*. Fear of unknown situations. In countries with high avoidance of uncertainty, there is an emotional need for laws that regulate action at all times (mainly Orientals). In countries with low avoidance, there is a significant rejection of formal rules (mostly Western ones). It would be expected that in a culture with high avoidance of uncertainty, managerial leadership will emerge, otherwise much anxiety would be generated. In countries with low avoidance of uncertainty, a too directive leadership would awaken strong resistance, because people expect to participate in decisions and feel inclined to assume responsibilities.

The Globe Project (Barata and Ripoll 2011; Gil et al. 2011) attempts to determine exactly these differences, but its scope exceeds our study. Nonetheless, the interested reader will find the material worthy of their time and attention. In this manual these issues will be addressed in Section 3.6: Managing Diversity: Multicultural Crews.

2.1.17   LEADERSHIP AND GENDER

Traditionally, women have not attained Officer status in the Merchant Marine. Fortunately, Western countries have observed increasing numbers of women in positions of responsibility and the nautical institutes now have more female students among their scholars. Even so, women in the merchant navy do not break the *glass ceiling* (the invisible barrier that prevents women from moving up in certain organizations). There is scarce literature on this subject because until recently it was not considered a subject worth inquiring about (Dragomir and Surugiu 2013; Magramo and Eler 2012; Thomas 2004).

There is also no agreement on whether women and men adopt different styles of leadership. In a meta-analysis (Van Engen and Willemsen 2004) investigating gender differences in leadership, it was found that (in laboratory contexts) there is a stereotyped perception of gender leadership: women are more oriented toward relationships and men are more oriented toward the task. In real contexts, however, these differences disappear.

The results show evidence of gender differences in leadership behavior, demonstrating that women tend to use more democratic and transformational leadership styles than men, while there are no gender differences in using other styles of leadership. Gender differences in leadership styles depend on the context in which male and female leaders work, as both the type of organization and the context are moderators of gender differences between leadership styles.

Regarding the effectiveness of leadership, the results of Barata and Ripoll's (2011) study show that there is no dissimilarity in team effectiveness when gender is considered, since the differences in the effectiveness of a team led by men or women are not significant. Finally, it was discovered that the styles of transactional and transformational leadership are significant to explain work team effectiveness in the case of women, but only transformational leadership is relevant in the case of men.

The ambiguity of the results may be due to the importance of the social context. Thus, in traditional organizations (bureaucratic, rigid, and markedly masculine), gender differences are masked because women adapt to the more masculine standards dominant in these types of organizations. On the contrary, in non-traditional organizations, women are free to exhibit their true style of leadership. What is clear is that women are no less effective than men when it comes to exercising leadership.

Gender studies are currently one of the much researched topics in the social psychology field, far exceeding the limits of this manual, in addition to the political and transcultural implications that surround this whole topic.

2.1.18   ETHICAL LEADERSHIP AND AUTHENTIC LEADERSHIP

Brown, Treviño, and Harrison (2005) have suggested the term *ethical leadership* to designate a more ethical and honest leadership style in a leader's relationships with his followers; probably a more modern and democratic style of concept understanding. This type of leadership is linked to the leader's perception of honesty and predicts the leader's effectiveness and the satisfaction and performance of his employees.

Another group of authors has proposed the concept of *authentic leadership* (Walumbwa et al. 2008), with the following characteristics:

- *Self-awareness* is the self-knowledge that a leader has of his own weaknesses and strength.
- *Transparency in relationships* will generate more trust among his followers if the leader is authentic.
- *Balanced processing of information* is the leader's objectivity and ability to access all information before making a decision.
- *Internalized morality*, the leader acts under his moral code, values, and principles that govern his behavior over social pressures or other groups' demands.

Recent research on this style of leadership shows a relationship with job satisfaction, commitment to the organization, and improved performance.

The landscape of leadership research is extensive and the reader can easily feel overwhelmed, but the important thing from a practical point of view, as officers in charge of a crew, is to draw conclusions about the leadership behaviors that most fit in each situation and adapt your conduct to a specific context. The many approaches discussed in this manual serve to make the reader reflect and extract the ideas that are common to all approaches to leadership, and apply them in their own context, because every officer is expected to be a leader.

2.2 THE EVIL LEADERSHIP: THE CASE OF *MARIA M*

We can find numerous examples of weak leadership leading to a maritime disaster. One of the most prominent is the famous case of the *Titanic*, where Captain Smith's lack of leadership when when he failed to apply technical criteria on navigation and security to the wishes of its owner, led to the well-known tragedy.

Studying more recent cases, we will analyze the case of the Italian-registered chemical tanker *Maria M*, which was sailing from Ventspils in Latvia on the afternoon of July 12, 2009, after having loaded 32,910 mt of diesel, and with a draft of 10.5 m. The captain was making his first shift on the vessel, and the outgoing captain was only with him less than two days. On July 14 at 21:50, the captain entered the bridge. At that time the ship was 7 nm from the anchoring area to take bunker in Gothenburg. After a series of maneuvers by the captain, the ship went off the planned course, and finally went aground.

The Swedish Investigation Commission has not fully clarified the reasons for the grounding, but all evidence seems to indicate that the captain was not sufficiently familiar with the instrumentation of the vessel, and that he was using the steer prior to the anchoring to verify the behavior of the ship and the response to orders at the helm. Research also points out how age and cultural differences played an essential role in cooperation on the bridge.

2.2.1   THE CREW

The crew was composed of three deck officers, a chief engineer and two officers, five deck crew, two engine crew, and another seven "unspecified" crew. The captain, the deck cadet, and the first engineer were Italian, the rest were Philippines. The 66-year-old captain had more than 20 years' experience on oil tankers; however, he only had two days with the outgoing captain to familiarize himself with a ship he had never been on before. The investigation also cites that the captain had participated in several bridge resource management (BRM) courses. The officer on watch at the time of the grounding was the third officer, 31 years old, and it was his first shift as a pilot. The 28-year-old helmsman had been on board for five months. This is the typical situation of any merchant ship today, with crews employed by an external contractor (in this case, Greek) and have little or no connection with the ship or with each other. The report also notes that fatigue does not seem to have caused the accident.

2.2.2   THE EVENTS

Navigation toward the anchorage area in Gothenburg seemed to proceed normally, until the captain made an appearance on the bridge. From that moment, the autopilot was disconnected and contradictory orders were followed in short time lapses, e.g., from 10° to starboard, followed by 10° to port, every few seconds. Seven minutes after entering the bridge, the captain took the helm himself and there were two consecutive hard turns of 30°, one to each side. At this point, the captain found himself discussing these maneuvers with the officer on duty, and even with the chief mate. He even called the third officer an idiot. Three minutes later, the helmsman returned to take over the helm. Later, a series of maneuvers turning to each band and with the engine that took to the ship to a zone of little draft that was not known to the crew. Soon, the first officer informed the bridge from the bow that they had gone aground and that ballast tank No. 1 was leaking (the crew hid this information from the Maritime Rescue Coordination Center [MRCC]). After several movements using the machine, the ship was refloated.

2.2.3   HUMAN FACTOR

All evidence points to the fact that the captain, although very experienced on other ships, was not sufficiently familiar with the *Maria M*. Additionally, the third officer of Philippine nationality had only been sailing as a pilot for three weeks, but he did have a good knowledge of the ship.

The investigators had the impression that the captain was an authoritarian person and at least twice he was heard calling someone an idiot on the bridge. The attitude of the third officer did not seem to be intrusive, rather submissive under the authority and greater experience of the captain, even though the third officer had better knowledge of the ship. It was clear that the Italian captain completely ignored the observations made by his Philippine officers.

The Swedish maritime administration concluded that the attitude of the captain and his lack of knowledge of the ship were the cause of the accident. Problems were also detected with the captain's deficient use of English and lack of communication with his officers.

On many occasions, the confusion of authoritarianism with strong leadership has led to catastrophic consequences. This accident could have been avoided if the captain had demonstrated real leadership, listening to his subordinates and dealing with them respectfully. The first task of a leader is to know his team and his equipment, better than anyone. Leaders must also be able to overcome and deal with the cultural and language barriers on board and follow established procedures. A true leader is not above norms and protocols, but sets an example in their fulfillment.

The interested reader can examine the report of this accident on the Swedish maritime administration website (Transport Styrelsen 2009).

2.3 LEADERSHIP IN EMERGENCY SITUATIONS

Natural leadership emerges when emergency situations occur on board. All the technical preparation in case of an accident is useless if an officer with a team (or a whole ship) is unable to manage emotions to avoid a possible disaster. We recall that more than 85% of maritime accidents are directly caused or are aggravated by human factors, and most of these human errors are not due to insufficient technical knowledge, but to failures in the decision-making process under pressure, or not being able to direct a team during a disaster.

In emergencies, leaders often emerge spontaneously and can exercise total control over a group. In an emergency situation, behavior is determined by multiple elements that are not always rational, an aspect of which the officer must be cognizant, as are having the skills and knowledge to be a leader.

According to Boin (2005), to be a successful leader in an emergency requires five key tasks:

1. Acting with meaning: A leader has the responsibility to foresee a crisis and manage the process to eliminate any contributing factor that could be avoided.
2. Making decisions and coordinating their implementation. Leaders are responsible for making the final decisions and enforcing them, ensuring that the group is fully informed and that as many participants as possible are included.
3. Creation of meaning: Leaders are in the spotlight when directing people in the right direction. It is the leader's ultimate responsibility to motivate the community to believe that they will overcome the situation.
4. Summarize and finalize: The leader must keep the affected parties on track to end the crisis.
5. Learning: It is essential that the leader assesses the situation and extracts the lessons learned either through failures or successes.

Communication is the most important aspect of leadership during a crisis.

2.3.1   Decision-Making Under Pressure

In an emergency, we encode and decode large amounts of information, resulting in a plan and taking action against time. Because it is a dangerous situation, fear and anguish can block intellectual performance. If this happens, there are easy exercises to become familiar with stress situations.

2.3.1.1   Role of the Cerebral Hemispheres in Emergency Situations

One of the key challenges is teaching behavior control in emergency situations. Preparation and previous training are essential to prevent the effects that these situations may have on behavior, such as blockages, exaggerated reactions, panic or loss of control. This may appear obvious, but crews on board are firemen, mechanics, doctors, and nurses, and the importance of effective preparation cannot be underestimated.

There is agreement among the authors on the existence of structural and functional differences between each cerebral hemisphere. It has been established, for example, that the ability to recognize faces lies in the right occipital lobe and to recognize language lies in the left frontotemporal area. Also, it has been concluded that the left hemisphere is responsible for acting and directing our behavior in a logical, orderly, organized, and sequential way.

Therefore, when emphasizing the importance of prevention or when carrying out an exercise on board, behavior is determined mainly by the left hemisphere of the brain.

Despite all the training and actual participation in rescues and emergencies (as happens on rescue ships), there are still cases of people suffering mental blocks in real situations, even clearly erroneous decisions are observed when analyzed calmly.

At the moment of danger, our bodies shift from a state of indifference to a condition of maximum alert, experiencing various physiological changes, leading to physiological challenges. In these circumstances, our behavior is mainly determined by the right hemisphere, characterized by being emotional and analogical.

The instinct for survival is strongly rooted in our behavior for evolutionary reasons. Comparative psychology reports curious facts related to safety and survival. In the face of a threat, living beings respond in two ways: with aggression or escape in the case of the human being, these behaviors can be understood as "doing something" (right or wrong), or falling into a behavioral blockade shock; this type of behavior in man comes from his ancestors.

Many studies affirm that the left hemisphere processes the abstract, rational, conceptual, and propositional information from logical analytical processes (De Martino et al. 2006; Hellige 2001; Rossi 1977; TenHouten et al. 1976). In addition, the left hemisphere regulates or inhibits the anxiety that comes from the emotional content of the right hemisphere. However, the right hemisphere prevails in tasks regulating images, visual activity, manipulation of spatial relations, and simultaneously understanding the totality from fragments.

The first phase of emergency preparedness is theoretical, since knowledge of the ship and its systems provides the officer with security and confidence when making decisions. From the neuropsychological perspective, we are working with the left hemisphere.

But in the event that this is not enough, emergency drills are mandatory. An ideal emergency drill is a simulation using actors playing different roles, including wounded, dead, or hysterical. This type of training is closest to a true emergency. In this way, we will be educating our right hemisphere; accustoming our brain to situations that are exceptional, so that when they occur, they are less of a surprise.

Studies carried out by different authors show that as we become familiar with a dangerous object or situation, it loses its threatening qualities, which means that it has less of an impact. This is the well-known psychological principle of habituation.

We know that a primary human fear is the unknown. Teaching the brain's hemispheres "pressure training" means exposing the participants to an intense level of stress, such as that produced and generated by *pseudo-collaborators* (people inclined to intervene in danger, but with negative results) in a real emergency situation. Generally speaking, hesitations and tremors are observed at the beginning, but tranquility and self-control soon emerge.

Neuropsychologists and students of cerebral hemispheric activity talk about the *procedural* or *organic memory*, meaning that the more our body practices a behavior, the more it becomes automatic. Hence, the importance of evacuation, fire, and emergency exercises in general.

2.3.2 Fear: Psychological Reaction to an Emergency Situation

Fear is an innate reaction of an individual to a danger that can lead to a mental block or overreaction, which may endanger the rest of the crew or the ship itself. On many occasions, fear can be expressed through clumsiness, even in routine tasks. It is the officer's mission to know his crew and understand how each member handles their emotions and especially their reaction to emergency situations. Particularly important to know is how to handle emotions on passenger ships, where in addition to dealing with personnel familiar with the ship, officers must do so with people from outside the environment who understand that they are in a hostile situation.

Facing a danger, we produce physiological alterations, aimed at increasing the activation of the body and preventing possible physical and cognitive overexertion. However, these overcapacities may be positive or negative depending on the person's reaction to a dangerous situation. Only training can prevent mental blocks in these situations. Control of human behavior facing an emergency on board is as essential as the technical preparation.

2.3.2.1 States of Fear and Forms of Appearance

Millán (2014) documented six states of fear that we can apply to a person facing danger. We can add one more: the "0," which refers to the person who does not feel fear, and is thus untrustworthy in an emergency. This type of person can generate other risks, to demonstrate their capabilities or by pure exhibitionism, so that they can aggravate the dangerous situation. The states of fear are

1. *Fear as prudence.* The person is aware of the danger and any action taken will be controlled psychologically by the subject.

2. *Fear as a precaution.* The person takes all measures in face of danger. He maintains a calm attitude and his behavior is appropriate to the situation.
3. *Fear as prevention or warning.* The person is nervous and hesitant. This state of fear presents disadvantages for the tasks inherent in the performance.
4. *Fear as acute anguish.* The person loses control of himself and generates a new emergency. He must be attended to immediately and removed from the area because his behavior could start a panic.
5. *Fear as fright.* The person is dominated by reflex acts, practically incapable of taking any appropriate action in face of the situation.
6. *Fear as terror.* The person is paralyzed by fear, unable to realize what is happening. They usually experience sphincter relaxation. Additionally, an increase in adrenaline discharge can have lethal effects. Terror can cause death due to a heart attack.

Fear is contagious and automatically transmitted. Words are not needed for this to happen, just a glance or a slight gesture is enough to spread fear. The dynamics of fear are different once it becomes collective fear. It changes the mental representation that each person makes of the world, that is, the same perception of reality. On many occasions, this distorted perception of reality can cause an emergency situation that did not previously exist.

Imagine a situation of abandoning a ferry with the emergency under control and without risk to the passengers. If a person suddenly panics thinking the ship is about to sink and dives into the water, a new emergency will be added to the existing one. When facing a situation of danger, the first thing to occur is the perception of the situation, but concurrently its evaluation. The assessment aims for survival. There is a conservation instinct that will trigger the escape/approach reaction, depending on the point stimulated in the hypothalamus.

There is abundant research on fear, but little on how to control it, except how our system manages it. Sweat, heartbeat, etc., can be modified by successive approaches to fear situations, simulating these circumstances through training.

Crew management in emergencies is the basis for mastering the situation. It is essential that, in an emergency, a minimum number of personnel be assigned to each task. A concentration of impassive people can aggravate the perception of risk. In a study published by the University of Marburg (Bierhoff, Klein, and Kramp 1991), willingness to help decreases as more people are present, due to the reduction in responsibility; therefore, the roles of the crew must be perfectly clear, and if the team is multinational, this factor must be considered when transmitting orders. It is a great advantage to have the roles of staff previously distributed, this way the person can mentally organize the tasks assigned. The images of a situation constitute, in a certain way, the situation itself. The most important work of the official is coordination and management, not physical intervention in the emergency.

2.3.2.2 Resilience (Ability to Overcome Adversity)

The concept of resilience, defined as the ability to overcome adversity and emerge stronger from it, improve from it, is key in the whole study of leadership. The main characteristic of the resilient person is the ability to resist adversity and coexist with risk factors without getting involved, distancing themselves emotionally with the necessary defensive elements. It is a feature not only necessary in emergency situations, but also in everyday life. In fact, this trait is as important as leadership for seamen, and it deserves a whole guide by itself.

Although there is little research on resilience in seafaring, it is worth citing the work by Doyle et al. (2016), as they reported that resilience increases the ability to overcome stress on board.

There are quite a number of works about resilience in other fields, particularly in the enterprise leadership field. Folkman (n.d.) has a web page full of resources for the interested reader, but some of them are not easily applicable to our environment, since the structure of a ship's organization is well established, as we know, and the roles are quite definite.

I strongly recommend theAmerican Psychological Association (2018) web page to find more resources to improve and promote resilience. Some key aspects to highlight are

- Significant relationships are essential for resilience. Try to build strong relations.
- Plan your actions in a realistic way. Do not bite off more than you can chew.
- Self-confidence and self-awareness must always be borne in mind. Realize who you are and your true abilities.
- Communicate effectively.
- Control your impulses. Think twice before acting.

Tips on building a resilient personality:

- Place your locus of control internally, that is, try to explain your success by means of your own skill and experience, instead of through luck or the help of others; and see failures as opportunities.
- A good sense of confidence and efficacy is a *must*. Try to be confident in your abilities and work capacity.
- Manage your own emotions. Try to be calm and rational in all situations. *Never run in the wheelhouse*, is a piece of advice that I heard the first day I went aboard.
- Grow and share positive emotions, as they are as contagious as the negatives.
- Set achievable goals, try to be ambitious, yet realistic.
- Be grateful for others' help and opinion, recognize their work as a means to build strong relationships.
- Be flexible in adapting to any new situation; a new ship or a new captain can be a challenging situation, and an officer must be able to adapt to new environments easily.
- *Assertiveness* is the way to express your thoughts and needs in a prosocial manner. This trait is key to growing your resilience on board, as we must keep the chain of command, but at the same time, express opinions and set limits on other's requirements. Try to be positively assertive, and never aggressive in manner and attitude.

- Assertiveness is not expressed in the same way across different cultures, as many have a long distance of power. Some national cultures may prefer people to be passive and might view assertive behavior as rude or even offensive, so try to be flexible and adapt to the ship's culture and the cultural background of the rest of the officers, mainly the captain.

2.3.3   Reactions in Emergency Situations

In emergency situations, each person's psychological reaction is different. This is primarily due to the complexity of our behavior.

In group emergencies, people's traits are enhanced or exacerbated, but there are also characteristics of the mass itself. Along with a loss of reasoning, responsibility diminishes and social limits tend to disappear. For example, psychopathic behaviors may appear in emergency situations. They can also produce an alteration of perception due to a change in the intentionality of consciousness—we see what we want to see, and we believe what we want to believe.

In an emergency with passengers, spontaneous leaders usually emerge from among the passengers. If we are able to detect them, they can be used to control other passengers. Otherwise, these people could turn against us, as they could motivate others to act against our interests, aggravating the situation.

It will be the officer's mission to detect and manage these innate leaders. Special attention should be paid to people who speak most, especially if they tend to challenge or discuss the orders of the crew; but also, those who isolate themselves from the rest, remaining silent and expressing no emotions, because they may be about to take radical action.

2.3.4   Verbal Communication in Emergency Situations

An emergency situation is the breakdown of harmony, a failure of security, and a challenge for the crew. The transmission of information must be verifiable to confirm that the sender said what he said and the receiver heard what he heard. Communication with foreigners deserves a special mention. It is important that there is a standard language on board that all crew members can understand, otherwise minimum instructions that everyone can follow without hesitation should be agreed upon during training.

In accordance with the STCW Convention and Code NGV 2000 (Code 2000), all crew members on board Ro-Ro and passenger ferries must be trained in crowd control, that is, the command and efficient evacuation of passengers in case of an emergency. Although the convention, as written, applies to Ro-Ro passenger ships only, the IMO recommends that the same rules should cover other passenger transport vessels. Most national administrations consequently also require certificates of public management training for crew members on board cruise ships and other passenger ships. One reason why the IMO, perhaps a bit hastily, added the crowd management training to the convention and the HSC Code may be the Ro-Ro passenger ferry MS *Estonia* disaster in the Baltic Sea on September 28, 1994.

2.3.5 CREW TRAINING FOR CROWD CONTROL

Crowd control is the ability to help passengers in emergency situations; for the control of passengers on stairs, in corridors and passageways, using procedures to prevent panic and the irrational behavior of others, and to communicate, instruct, and inform passengers. Other crew members will also mobilize passengers to help convince and calm passengers when the emergency is over. To deal with this, regulations require knowledge of crisis management and human behavior. This training focuses on applied psychology to be competent to evaluate the reactions of passengers and crew.

According to the STCW, the crew must have the capacity to supervise and manage a passenger in emergency situations.

- The team must offer passengers relevant information in an emergency situation. (Once people know why they are required to take certain actions they are more willing to do so.)
- The team must assume a leading role in an emergency situation. (Passengers must have faith in the crew through their uniform and their verbal and nonverbal behavior.)
- The team must react appropriately after evaluating the reaction of the passengers.
- The team must mobilize some passengers to help.
- The team must convince the passengers that the emergency situation is more serious than it appears, if this is the case.

2.3.5.1 Chaos Is Inherent in All Emergencies

The reasons for the IMO regulations are the lessons learned from the past, but probably also that passengers often, by far, outnumber crew members. This is less obvious in the case of passenger ships, but there are cases on ferry ships of crews of 50 people transporting 2000 passengers. Facing any emergency situation, with all its inherent possibilities for chaos, the crew needs to be intelligent, effective, and exhibit competent and trained behavior to handle the situation.

Normally, evacuation plans and simulations are intended to cover all conceivable situations and are the crew's guide to dealing with disaster situations in an organized manner. These plans will sometimes have to be adapted to situations that are difficult to imagine. Plans and simulations, therefore, have limitations.

2.3.5.2 Training for Crowd Control

During drills, officers should include practice to train crews to deal with groups of people, with an emphasis on distinguishing the most vulnerable (children, the elderly, the disabled), as well as detecting potential breakers in the group, that is, those who tend to discuss or disobey the orders of the crew or constantly demand information. The crew should be instructed to communicate these observations to an officer as soon as possible, but without interfering with operational procedures.

2.3.5.3   Special Emphasis on Information for Officers

The officer in emergencies with passengers should focus on handling information: most passengers complain about a lack of information; therefore, it is important to transmit the information required at any time, without hiding data, which can cause a loss of prestige to the company. In the event of a failure of either equipment or organization, it is preferable to apologize for the failure and admit guilt rather than trying to deny it, which could put the passenger against us.

Publishing inaccurate information through the public address system, communicating insufficient information and poor reporting can create an uncontrollable situation for the rest of the crew. Therefore, officers must pay attention to the various hazards and their expected consequences for the passengers and reactions to the crew. For this, a checklist should be prepared before issuing any announcement over the public address system.

2.3.6   How to Inform Passengers in Case of Emergency

- Information to passengers takes high priority.
- Be prepared for the stress reactions of passengers and crew members.
- Control information.
- The crew must be better informed than the passengers.
- Never lose credibility.
- Introduce yourself (title and name) and keep your voice under control.
- Speak slowly and clearly. Inform without being talkative. Avoid technical language (passengers might think that the crew is trying to hide information with the use of incomprehensible jargon).
- Messages must be in accordance with events the passengers themselves are experiencing.
- Report everything passengers can perceive or understand.
- Do not hide the real risks.
- Passengers want truthful information.
  - Tell passengers what has happened.
  - Offer a personal assessment of the situation, this will convey closeness.
  - Give details on what measures have been taken as this will convey confidence that the crew is trying to solve the problem using every means.
  - State the expected result, hope must be instilled.
  - Indicate if the situation has been reported ashore and that assistance is on the way.
  - Point out that the team is trained for the task and that passengers must follow the instructions of the crew at all times.
  - Announce the time of the next situation report.
  - Give information frequently. Every 10–15 minutes during an active phase. Every half hour to an hour during a phase of greater stability.
  - Limit instructions and information.
  - Specify the precise time for the next call, and be punctual.
  - Repeat the above information and also provide it in written form if possible.

- Uncertainty is the worst state.
- Notify about meeting points and where people who have lost a family member should go. Usually, the most stressful situations for passengers are those of people lost on the ship, especially children. These people should be given a point to go to and priority help.

2.3.6.1 Alarm Signals Are Not Enough

On many occasions, the passenger does not understand the alarm signals, and is unable to distinguish if there is a real emergency or if it is a simulation; therefore, we must use all means at our disposal to prevent the passenger the passenger from taking actions which could endanger his own safety or the safety of others, even beyond those indicated by the protocols if necessary. This requires using voice messages in languages necessary to reach all people on board.

In the event of an alarm signal, the normal reaction is that passengers will stop the activities they are engaged in, and many will start looking for the device that produces the alarm signal. After a few moments of silence, many will begin to discuss the signal, investigating possible interpretations, discarding misunderstandings, and reaching a conclusion regarding its meaning. This slower reaction pattern is formed from a mixture of surprise and disbelief, along with reluctance to believe that something stressful and potentially dangerous is occurring in the midst of a pleasant activity.

The alarm signal, as such, does not offer passengers any guidance about what they should do. Experience shows most passengers do not see themselves reflected on or affected by signaling and if they do, they cannot remember the placard or their location. Experience also shows that, in general, the emergency alarm signal is too complicated. Most do not instantly consider that seven short (or more) and one long are a warning signal.

Passengers may not believe that five or six short warnings followed by a long one indicate a general emergency alarm. In addition, the signal does not have a rhythmic pattern or melody unlike other well-known signals, including fire and ambulance signals.

In addition to this, the general emergency alarm warning is not easy to distinguish for people who are unable to count, such as young children. Using only alarm signals, passengers are left alarmed and confused. The same alarm signal can create a disorganized situation if people evaluate the situation differently, following one another in various actions. Hence, the need for verbal information.

A voice message can provide passengers with directives ensuring that the passengers' perception of the situation and the options for actions are in line with the reality and the objectives sought by the command team. Providing directives, instructions, and information through the public address system is the most efficient way to handle passengers.

2.3.6.2 Responsibility Increases the Potential for Action

It is also important to emphasize the behavior of the crew. A sense of responsibility is important because it is an established psychological fact that increasing the sense of responsibility increases a person's ability to withstand stress. Crew members must always be aware of their responsibilities to passengers in emergency situations. This awareness, in combination with their training, will help them overcome their own emotional turmoil and stress, and react and act in a more rational way.

We can also observe that passengers' sense of responsibility makes them volunteers, as previously described. We should take advantage of these group members because some passengers might find it easier to identify with them than with a crew member who has more knowledge of the ship. These spontaneous volunteers can rise up for two main reasons. First, because they feel they can help, and secondly because having a role and a responsibility helps them overcome their own stress. Doctors, nurses, police, marines, and firefighters along with other professionals are often especially prone to dealing with their stress in this constructive manner, and they should be welcomed by any small team as valuable reinforcement.

2.3.6.3   Authority Is a Prerequisite in Crowd Control

The vast majority of the crew on a passenger ship are steward personnel, so they are not specially trained in crowd control and therefore lack inherent authority. They also work in conditions on board that are not commonly considered superior and their uniforms cannot be designed to indicate authority.

In an emergency situation, the catering crew constitutes most of the emergency team and is expected to be in charge of the passengers. This crew should be trained to instill authority in emergencies, project their voice, and know how to use a megaphone. Consequently, they learn to give clear and concise instructions and to be patient with passengers who do not understand or do not follow them. They must also be instructed to inform officers of any news emanating from their group.

Another aspect to consider is the information channel to the crew, since the one that addresses the crew cannot be the same one that addresses the passengers. Particularly difficult situations can arise when the noise level in an open area is high, and a member of the crew is unable to hear messages through a speaker when passengers can hear them. This situation can cause misinterpretations.

A crew member must never appear confused or uninformed to the passengers, as they would rapidly lose their authority. If something is not known, it is better to admit it, and then ensure that passengers are informed as soon as the information becomes available. They must, therefore, be equipped with radio portable telephones and their own information channel. The portable VHF is not intended to be used at all times because it also operates as a symbol of authority, and using a headset gives passengers the impression that the crew is better informed.

2.4   CATASTROPHE OF THE *ESTONIA*

The disaster of the *Estonia* has all the distinctive features of an unthinkable situation and represents in every detail the worst possible scenario. The accident occurred in Estonia in the middle of the night and in stormy weather. From the start of the accident at 01:15, until the ferry disappeared below the surface at 01:55, only 40 minutes elapsed. Most of the passengers and crew members were asleep, and were awakened by metallic blows followed by a list (tilt) of around 5°. From the start of the accident until the list was too serious to evacuate the crew, only about 15 minutes passed.

The final report can be found at http://onse.fi/estonia/.

2.4.1 THE CREW

The crew comprised 186 people, including the master. There were five deck officers, a radio officer, and eight ratings. There was also a doctor. The engine department comprised eight engineers plus eight ratings. There were 8 officers in the catering area and 113 crew members.

The work system was very good for the crew, accounting for two work weeks, followed by two out. At the time of the accident, the crew was in their thirteenth working day. All deck officers and most of the crew were Estonian. The working language on board was Estonian, and was understood by all crew members. Everybody had the required certification for their position on board.

Both the training program and the safety management were revised shortly before the accident. All the emergency and evacuation groups were clearly stated in the safety plans. They trained every two weeks and the equipment was properly managed and maintained.

2.4.2 THE EVENTS

The ship left Tallinn in normal full operating conditions, with good visibility and rain showers. One second officer and a third mate were on watch that night. The vessel had a moderate list to starboard due to cargo and wind conditions. At midnight, weather conditions worsened with increasing wind speeds of 15–20 m/s and waves of 3–4 m. Fin stabilizers were in use, but most of the passengers were seasick.

Around 01:00, a crew member heard a metallic clunk at the bow. He informed the bridge and was instructed to investigate where the noise came from. Soon after, other noises were heard coming from the bow and the officers on watch sent another able seaman to the bow.

At 01:15, the visor tilted and separated from the bow. The ramp opened completely and water from the sea entered the car decks, quickly listing the vessel to starboard. At 01:30, the list was almost 90°. At about 01:50, the ship sank.

Passengers panicked and rushed to reach the lifeboats, but most of the passengers were trapped in their cabins by the water. Life jackets were provided to those who managed to reach the deck. The situation degenerated into total chaos.

About 300 people arrived on the outer deck, of which 137 survived. Most of the deceased were in various states of undress. Those who managed to get on the life rafts were adrift in the storm until dawn, between four and seven hours before being rescued. Many perished during these hours from exhaustion and cold. The water temperature was 12°C and the air temperature was only 9°C. Eight hundred and fifty-two people lost their lives.

2.4.3   THE HUMAN ELEMENT

From the survivors' statements we learned about the activities of the crew members and the reaction patterns of the crew and passengers. There were reports of panic, paralysis, shock, inability to understand, the impossibility of finding decisive action, exhaustion, and other altruistic and heroic behaviors.

The report highlighted the crew's actions, particularly their failing to reduce speed before investigating the noises, and their being unaware that the list was being caused by water entering the main garage.

The only alarm procedures followed took place at about 01:20 when a weak female voice called in Estonian "Alarm, alarm, there is an alarm on the ship," over the public address system, which was followed immediately by an internal alarm for the crew (useless for passengers), then one minute later by the general abandon ship alarm. Also, the passivity of the crew and their delay in responding were criticized by investigators.

We should emphasize that failures in communication and alarm signals were vital in keeping the passengers from reaching the deck. The ferry quickly listed by 30°, so all attempts to move inside the ship were futile. If the warning signals had been understandable and highlighted, many more lives would have been saved. Initially, the only message that was issued was not clear enough and was not repeated sufficient times. Some crew members, acting on their own initiative, organized the evacuation locally by guiding passengers, helping and arranging groups, and releasing life rafts. There were reports of very few passengers taking action to organize and direct others to evacuate—natural leaders. Survivors were mainly those who kept calm and retained their self-control to reach the life rafts. Those who panicked were lost. On the other hand, some people struggled to survive at others' expenses, showing irrational behavior.

The commission stated that strict orders should have been given from the bridge through the public address to passengers and crew. Language was also a problem in the entire emergency management, shifting from Estonian to Finnish and English when sending distress calls over the radio, delaying rescue operations, and between passengers and crew.

The passengers did not know how to use the survival equipment, there were failures in coordination by the crew, leaving many passengers alone, and the emergency signals were sometimes not understandable. The atmospheric conditions were very adverse and the rafts did not have an adequate design. These failures had not been detected in previous exercises or inspections. Also, the rescue operation wasn't properly managed, for instance, a single rescue helicopter was not enough and many of the rescuers were injured.

The main conclusion is that training is the best tool for officers and crew to deal with emergencies and that nothing should be left to improvisation.

REFERENCES

Allport, Gordon W. 1954. "The Historical Background of Modern Social Psychology." In Lindzey Gardner (ed.) *The Handbook of Social Psychology* (pp. 3–56). Addison-Wesley.
American Psychological Association. 2018. "*The Road to Resilience.*" APA. https://www.apa.org/helpcenter/road-resilience.aspx.

Barata, Meirijane Anastácio, and Margarita Martí Ripoll. 2011. "El Liderazgo en una Perspectiva Internacional: Un Estúdio Comparativo Entre Líderes Brasileños y Españoles de Acuerdo con el Proyecto Globe." *Revista de Psicologia* 2 (1): 35–49.

Bass, Bernard M. 1985. *Leadership and Performance beyond Expectations*. Collier Macmillan.

Bass, Bernard M, and Ralph Melvin Stogdill. 1990. *Bass & Stogdill's Handbook of Leadership: Theory, Research, and Managerial Applications*. Simon & Schuster.

Bennis, Warren G. 2001. *Conducir Gente Es Tan Difícil Como Arrear Gatos:¿Los Líderes Se Pueden Hacer?* Ediciones Granica SA.

Berson, Yair, and Jonathan D Linton. 2005. "Leadership Style and Quality Climate Perceptions: Contrasting Project vs. Process Environments." *International Journal of Technology Management* 33 (1): 92–110.

Bierhoff, Hans Werner, Renate Klein, and Peter Kramp. 1991. "Evidence for the Altruistic Personality from Data on Accident Research." *Journal of Personality* 59 (2): 263–280.

Boin, Arjen. 2005. *The Politics of Crisis Management: Public Leadership under Pressure*. Cambridge University Press.

Brown, Michael E, Linda K Treviño, and David A Harrison. 2005. "Ethical Leadership: A Social Learning Perspective for Construct Development and Testing." *Organizational Behavior and Human Decision Processes* 97 (2): 117–134.

Burns, James M. 1978. "*Leadership New York.*" Harper & Row.

Castro Solano, Alejandro. 2006. "Teorías Implícitas Del Liderazgo, Contexto y Capacidad de Conducción." *Anales de Psicología* 22 (1): 89–97.

Code, High-Speed Craft. 2000. *International Code of Safety for High-Speed Craft*. International Maritime Organization.

Doyle, Niamh, Malcolm MacLachlan, Alistair Fraser, Ralf Stilz, Karlien Lismont, Henriette Cox, and Joanne McVeigh. 2016. "Resilience and Well-Being amongst Seafarers: Cross-Sectional Study of Crew across 51 Ships." *International Archives of Occupational and Environmental Health* 89 (2): 199–209.

Dragomir, Cristina, and Felicia Surugiu. 2013. "Seafarer Women-Perception of the Seafaring Career." In *Proceedings of the Second International Conference on Economics, Political, and Law Science (EPLS'13): Advances in Fiscal, Political and Law Science*, Brasov, Romania, 1–13 June, 2013 (pp. 15–18).

Edwards, Bryan D, Suzanne T Bell, Winfred Arthur Jr, and Arlette D Decuir. 2008. "Relationships between Facets of Job Satisfaction and Task and Contextual Performance." *Applied Psychology* 57 (3): 441–465.

Engen, Marloes L Van, and Tineke M Willemsen. 2004. "Sex and Leadership Styles: A Meta-Analysis of Research Published in the 1990s." *Psychological Reports* 94 (1): 3–18.

Fiedler, Fred E. 2006. "The Contingency Model: H Theory of Leadership Effectiveness." In John M Levine and Richard L Moreland (eds) *Small Groups: Key Readings* (pp. 369–382). Psychology Press.

Folkman, Joseph. 2018. "Joseph Folkman." Accessed December 24, 2018. https://www.forbes.com/sites/joefolkman/#7864c45f2fad.

Gil, Francisco, Carlos Alcover, Ramón Rico, and Miriam Sánchez-Manzanares. 2011. "Nuevas Formas de Liderazgo en Equipos de Trabajo." *Papeles del Psicólogo* 32 (1): 3–13.

Hater, John J, and Bernard M Bass. 1988. "Superiors' Evaluations and Subordinates' Perceptions of Transformational and Transactional Leadership." *Journal of Applied Psychology* 73 (4): 695–702.

Hellige, Joseph B. 2001. *Hemispheric Asymmetry: What's Right and What's Left*. Vol. 6. Harvard University Press.

Hellriegel, Don, and John W Slocum. 1998. *Administración*. International Thomsom Editors.

Hersey, Paul, Kenneth H Blanchard, and Walter E Natemeyer. 1979. "Situational Leadership, Perception, and the Impact of Power." *Group and Organization Studies* 4 (4): 418–428.

Hilton, James L, and William Von Hippel. 1996. "Stereotypes." *Annual Review of Psychology* 47 (1): 237–271.

Hofstede, Geert. 1983. "The Cultural Relativity of Organizational Practices and Theories." *Journal of International Business Studies* 14 (2): 75–89.

Hollander, Edwin P. 1985. "Leadership and Power." In Lindzey Gardner and Elliot Aronson (eds) The Handbook of Social Psychology (pp. 485–537). Random House.

STCW, IMO. "Standards Training." *Certification and Watchkeeping, IMO Publication, London* (2010).

Lepore, Lorella, and Rupert Brown. 1997. "Category and Stereotype Activation: Is Prejudice Inevitable?" *Journal of Personality and Social Psychology* 72 (2): 275–287.

Lewin, Kurt. 1939. "Field Theory and Experiment in Social Psychology: Concepts and Methods." *American Journal of Sociology* 44 (6): 868–896.

Magramo, M, and G Eler. 2012. "Women Seafarers: Solution to Shortage of Competent Officers?" *International Journal on Marine Navigation and Safety of Sea Transportation* 6 (3): 397–400.

Martino, Benedetto De, Dharshan Kumaran, Ben Seymour, and Raymond J Dolan. 2006. "Frames, Biases, and Rational Decision-Making in the Human Brain." *Science* 313 (5787): 684–687.

Millán, Juan M Fernández. 2014. *Gestión e Intervención Psicológica en Emergencias y Catástrofes.* Ediciones Pirámide.

Nisbet, Richard E, and Lee Ross. 1980. *Human Inference: Strategies and Shortcomings.* Prentice-Hall.

Omar, Alicia. 2011. "Liderazgo Transformador y Satisfacción Laboral: El Rol de la Confianza en el Supervisor." *Liberabit* 17 (2): 129–138.

Paris, Laura, and Alicia Omar. 2008. "Predictors of Job Satisfaction among Physicians and Nurses." *Estudos de Psicologia (Natal)* 13 (3): 233–244.

Pettigrew, Thomas F, and Roel W Meertens. 1995. "Subtle and Blatant Prejudice in Western Europe." *European Journal of Social Psychology* 25 (1): 57–75.

Rosch, Eleanor. 1999. "Principles of Categorization." In E Margolis and S Laurence (eds) *Concepts: Core Readings* (pp. 189–206). MIT Press.

Rossi, Ernest. 1977. "The Cerebral Hemispheres in Analytical Psychology." *Journal of Analytical Psychology* 22 (1): 32–51.

Shamir, Boas, Robert J House, and Michael B Arthur. 1993. "The Motivational Effects of Charismatic Leadership: A Self-Concept Based Theory." *Organization Science* 4 (4): 577–594.

Sherif, Muzafer. 2010. *The Robbers Cave Experiment: InterGroup Conflict and Cooperation.* (Originally published as *Intergroup Conflict and Group Relations*). Wesleyan University Press.

Solano, Alejandro Castro, A Perugini, M L Benatuil, D Nader, and M C Solano. 2007. *Teoría y Evaluación Del Liderazgo.* Paidós.

TenHouten, Warren, John W Morrison, E Paul Durrenberger, and S I Korolev. 1976. *"More on Split-Brain Research, Culture, and Cognition."* University of Chicago Press.

Thomas, Michelle. 2004. "'Get Yourself a Proper Job Girlie!': Recruitment, Retention and Women Seafarers." *Maritime Policy and Management* 31 (4): 309–318.

Transport Styrelsen. 2009. "Chemical Tanker MARIA M." www.transportstyrelsen.se.

Vroom, Victor H, and Philip W Yetton. 1973. *Leadership and Decision-Making.* Vol. 110. University of Pittsburgh Press.

Walumbwa, Fred O, Bruce J Avolio, William L Gardner, Tara S Wernsing, and Suzanne J Peterson. 2008. "Authentic Leadership: Development and Validation of a Theory-Based Measure." *Journal of Management* 34 (1): 89–126.

Wofford, J C, Vicki L Goodwin, and J Lee Whittington. 1998. "A Field Study of a Cognitive Approach to Understanding Transformational and Transactional Leadership." *The Leadership Quarterly* 9 (1): 55–84.

Yukl, Gary, Angela Gordon, and Tom Taber. 2002. "A Hierarchical Taxonomy of Leadership Behavior: Integrating a Half Century of Behavior Research." *Journal of Leadership and Organizational Studies* 9 (1): 15–32.

3 Teamwork

Traditionally, an officer on a merchant ship works individually rather than in a team, as can be seen in other organizations. However, the growth and development of technology together with decreasing crew numbers, makes teamwork skills increasingly necessary, not only in dealing with other officers, but also with subordinates. It is also important to know how to form effective working groups and the processes to which they are subject. We have already discussed some of these issues under the prism of leadership, now we will approach teamwork in a more global way.

Teamwork is a competitive advantage for any organization, not only because of the synergies that are achieved by the integration of various personal skills, but also because participatory competition allows an increase in productivity, innovation, and job satisfaction (Aritzeta, Ayestaran, and Swailes 2005). Working as a team means not only meeting and carrying out an activity in common, but also creating a space where all members can give their best for the sake of a common goal.

We have to distinguish between a group and a team because they are different concepts. A group is the union of several individuals who may or may not have characteristics in common. A team, however, is the union of several people who have complementary skills and a common goal. All members are responsible for the team results and agree on common rules and values of internal functioning, whether they are tacit or explicit.

First of all, we will try to understand the processes and interactions between groups, and then we will begin to have a better understanding of the dynamics that govern teams.

3.1 GROUP STRUCTURE

The latter half of the twentieth century saw a significant increase in the study of groups from a psychological perspective. The nature of the work in the merchant navy requires working in groups and teams, whether on large or small vessels. Modern psychology offers us tools to better understand processes that can help avoid conflicts and increase satisfaction and productivity within and among groups.

Moreland and Levine's (1992) models on group composition are the most followed in the literature. The authors show ways to categorize groups according to individual characteristics, demographics, opinions and beliefs, perspectives, etc.

3.1.1 EFFECTS OF GROUP SIZE

The advantages and disadvantages of increasing the size of a group:
Advantages

- Greater availability of resources
- Ambitious goals

- Diversity
- Greater legitimacy

Disadvantages

- Coordination problems
- Loss of motivation
- More conflict and less cooperation
- Appearance of anti-normative and disruptive behaviors
- Less participation
- Less satisfaction

As we can see, the size of a group can sometimes have contradictory effects. On board ship, it is not unusual to work in small groups, with modern crews severely reduced due to automation, the exception being passenger ships, where the roles are inverted. We must also consider other variables, which we call moderators of the main effects, such as members' own capacities and their motivation.

3.1.2   Effects of Group Diversity

Crew diversity has always existed on ships, but has increased in modern times. We will discuss this in Section 3.6, but suffice to say that diversity, as mentioned in the previous section, has positive and negative nuances. It can improve relationships with members from different ethnic groups, create synergies, and favor flexibility and creativity. But it may also increase conflicts, reduce group cohesion, and increase the possibility that some members might want to leave a group due to insufficient integration.

3.1.3   Effects of the Combination of Individuals in the Group

Moreland and Levine (1992) point out that when the characteristics of individuals are combined in an additive way, the behaviors are independent of the group, and are more related to the personal traits of its components. The other combination rule is interactive, so that the group members are interdependent, and the person vary their behavior depending on the group.

3.1.4   The Concept of Group Structure

Several factors define the structure of a group, as we will see in more detail.

3.1.4.1   Intragroup Differentiation

Intragroup differentiation can come from the formal structure of the group, which is usually our case, where each crew member has established roles. However, there are categorization processes based on the prestige of the members and their contributions to the adjustment to group or organizational norms. Some authors (Terry and Hogg 1996) focus on specialization within the group in the intergroup context, that is, external. For them, when a group prototype is built, a process of depersonalization and uniformization of the individual takes place, homogenizing the behavior within the group. According to these authors, it occurs mostly in traditional groups and small organizations so it can affect our work context. Another perspective places more emphasis on social attraction, felt by other group members depending on whether they represent the prototype. Here, if the prototype is suitable for the organization (high productivity, good adjustment to standards, etc.) it obviously will be convenient to cultivate, or try to change it otherwise.

3.1.4.2 Hierarchies in Groups

Crews on merchant ships are subject to a hierarchical structure that is imposed by the shipowner or by tradition; however, another power structure exists, but hidden in plain sight, what we call the *status system*. This is a social system in which status is derived from a person's position or achievement.

In equivalent groups (as ours usually are), the most influential members and those considered more competent tend to be more involved. Compliance with group rules also seems to influence this hierarchy. Other theories focus on a group's expectations regarding the performance of each member; thus, the more a person contributes to the group task, the greater his status and the perception that he is more competent and creative.

Another approach places greater emphasis on dominance, and is based on each person's evaluation of others' appearance and verbal behavior. This dominance can be overt or subtle and greatly depends on members' ability to handle stress (resilience)—the more resistant, the less influenced by dominance. Other authors speak of mixed models that integrate diverse characteristics, and these are probably closest to the reality of a group.

3.1.5 GROUP REGULATIONS

This topic has been extensively investigated. We will attempt to summarize group regulations, focusing on the pertinent facts.

* Definition: There are several definitions for group regulations that can be grouped by saying that the norms of behavior are frames of reference that produce uniformity in members' behavior, or regulations on behavior establishing what is acceptable in a group.
* Functions of the rules: Rules have individual and social functions. Among the first is the creation of a reference framework that explains reality. In terms of social functions, rules coordinate the interactions of members, direct goals, and maintain group identity.
* Development of rules: Rules can be generated internally within a group, or externally, which is most commonly the case on a ship, where there is a certain regulatory rigidity about what is expected of each crew member. The internal origin of regulations can be from manuals that each member has of how to face situations and negotiate exchanges, or from existing initial patterns that have crystallized into the norm. Likewise, they can be imposed more or less explicitly by the group leader or derived from the prototypical or ideal member.

3.1.5.1   Effects of Deviation with Respect to the Group Norms

In general, deviations break the uniformity of members' behavior and give rise to a phenomenon that we must be more vigilant about on board ship, which is the *pressure toward uniformity* and the rejection of those who deviate from group cohesion. This phenomenon is also related to the group's desire to achieve consensus without which productivity, satisfaction, and the generation of creative solutions may be affected.

Another aspect is the introduction of *perverse rules*, which are norms that are impossible to comply with and generate corruption and demoralization.

3.1.5.2   The "Black Sheep" Effect

Studied mostly by Marques and Paez (1994), the "black sheep" effect deals with the denigration of group members who show undesirable characteristics or break the rules, comparing to the attitude generated by those same behaviors in members of other groups.

These strategies exist to eliminate members who contribute negatively to the creation of group identity, which seems to occur in almost all groups to a greater or lesser extent, depending on whether the norms are relevant (the effect is given) or irrelevant. However, the effect is the opposite when members outside the group are judged.

This effect also depends on the status of the deviant member—the greater the status, the more valued are the deviations from the norm; they are tolerated in a certain way. However, the effect of criticism is usually the opposite, with internal criticism better tolerated than that coming from outside the group.

3.1.6   Group Roles

Group roles can be defined as people's behavior within a group, which entails both rights and duties. In our context, the roles within groups will be formal and established hierarchically; however, there may be other roles of a more informal nature that arise spontaneously, and may be assigned by the group leader or other in the group.

Role differentiation within a group leads to goal achievement; however, a trend in our context is the extreme rigidity in certain roles that prevents functional communication within the organization—having to go through several filters before reaching the officer or captain—excessive rigidity is a danger to be avoided. However, role differentiation serves to organize group functioning and make it predictable, in addition to favoring the inclusion of members in the group, even members who deviate from the rules have an important role in the group dynamics, since they serve to fix rules, or they can produce changes in the group behavior. According to Benne and Sheats (1948), the types of roles within a group can be distinguished as follows.

Task roles

Coordinate and facilitate the solution to problems:

- Elaborator
- Activator

- Orientator
- Evaluator
- Gives information
- Asks for information
- Initiator
- Procedures expert

Construction and maintenance of the group:

- Seeks compromise

Support and regulate group-oriented attitudes:

- Animate
- Follower
- Harmonizer
- Observer
- Goalkeeper

Individual roles:

- Aggressor—*potentially dysfunctional*, they do not address the task or the maintenance of the group
- Blocker
- Seeker of recognition
- Dominant
- Evader

3.1.7 PROCESSES OF INFLUENCE INSIDE GROUPS

Merely belonging to a group modifies a person's thinking and behavior. These processes are known as group influence, and it is important to know them and how they can modify behavior, thereby avoiding the risks they imply.

Social influence is something that we cannot escape as members of society. It builds our behavior from childhood, according to the type of education; how a person can change their perception or opinion under group pressure, how we prejudice certain social groups or foreigners by the mere fact of being them, or how we act in socially tolerated situations including alcohol consumption (a great problem on board). We will analyze the main effects that groups exert on individuals, without losing the perspective and scope of this manual.

3.1.7.1 Conformism

Conformism is the modification that a person makes to their behavior, adapting it to the reference group. It can also be expressed as the normative influence of the rules and beliefs prevailing in a group.

3.1.7.2   Uncertainty and Consensus

Others' opinions and attitudes can influence an individual who is uncertain about a situation. People tend to avoid unknown or unusual situations due to the uncertainty they generate, by observing and imitating others' behavior, whether they are right or wrong.

Festinger (1950) considered that judgments, beliefs, opinions and attitudes must have a valid basis, thus, when an individual has to construct social reality he finds himself in a situation of uncertainty or ambiguity that he avoids looking at the actions of his reference group, and especially of influential people. People in general can make objective judgments about a physical reality, but in a social situation it is more difficult to make subjective assessments.

The problem is that for a response to be considered valid, it must be in accordance with the social consensus, meaning in accordance with the majority. In this way, there is a danger of building opinions, beliefs, and erroneous attitudes, because they are perceived not to be the predominant ones.

Sherif (1936) argued that individuals internalize social norms and act according to such norms, their influence being constant and lasting. Many of these social principles and conventions go unnoticed because they are so internalized that we fulfill them as a matter of course.

Asch and Guetzkow (1951) performed a classic experiment in group psychology, where they demonstrated how individuals gave answers objectively and evidently erroneously, by adapting to the majority opinion of the group. One version of this experiment that I do with my students is to ask a volunteer to leave the classroom and then agree with the rest of the class that a football player's shirt in a photograph being projected is red, when it is evidently orange. Then, the volunteer is brought back into the class and several people are asked about the color of the shirt. Everyone will affirm it is red, and then the volunteer is questioned about the color he observes. Experimental studies claim that more than 30% of people modify their responses to adapt to the mainstream.

3.1.7.3   Factors That Influence Conformism

Conformism increases with group control over the individuals in a group and with the interaction between them. Additionally, other factors provide broad empirical support:

- Greater when the answers are given in public than in private (Asch 1956).
- When the ties within the group are narrow (Lott and Lott 1961).
- When there is a common goal (Deutsch and Gerard 1955).
- In collectivist countries (Bond and Smith 1996).
- Women more than men (Eagly and Carli 1981).
- When the person is perceived as incompetent (Hochbaum 1954).
- When the difficulty or ambiguity of the task increases (Coleman, Blake, and Mouton 1958).
- Increases if there is a correct answer objectively, or can be verified empirically (Crutchfield 1955; Insko et al. 1983).
- Decreases with increasing involvement of the person in the task or opinion (Deutsch and Gerard 1955).

- Increases with relevance to the subject. Conformism is usually greater when there is a similarity between the source and the individual, although this effect mainly appears when it comes to making judgments of value (Jones and Gerard 1967).
- Increases with the size of the group, although the effect stabilizes with three members (Asch and Guetzkow 1951).
- It decreases with insufficient unanimity. Hence, it is enough for one member of the group to express an opinion contrary to the majority so that the pressure toward conformity decreases (although this opinion can be incorrect) (Asch 1956).

3.1.8 MINORITY INFLUENCE

Not only majorities but also certain minorities can exercise influence. In this way, it has been empirically demonstrated that, while majorities have clear influence, minorities can exert effects of converting some people or creating latent influence (Maass and Clark 1984). Thus, the ideas of radical groups have a certain social significance. Such influence can also be manifested in small groups. How minority groups gain influence is beyond the scope of this manual, since there is no consensus among researchers.

In any case, certain studies suggest that minorities have to generate conflict to obtain influence, so that the majority's confidence in their position is reduced, allowing the minority to force negotiation or action. These processes are important in understanding certain pressure groups on board ship, such as trade unions, and why they need conflict (Moscovici 1981). Moscovici defined five styles of behavior: involvement, autonomy, rigidity, consistency, and equity, with consistency and the style of negotiation of greatest importance. Consistency in behavior allows a minority to be perceived as an alternative. Negotiation style primarily refers to rhetoric or the way of negotiating, with rigidity characterized by extremism and intransigence, blocking negotiations, and rejecting conditions. Pressure groups use these mechanisms to obtain influence on individuals, and finally on the majority.

3.1.8.1 Mechanisms of Defense against Minority Influence

- *Psychologizing*: Psychologization consists of condemning or excusing individuals for their psychological traits, even if there is no proof of the existence of these traits.The individual explains the message of the minority according to their characteristic, thus, because these tend to have negative features, it will be inclined to scorn their positions.
- *Denial*: A focus and active rejection of minority positions. Paradoxically, it can increase knowledge of their position and, in this way, their influence.
- *Social cryptomnesia*: When the majority ends up appropriating some ideas of the minority, forgetting their origin.

In short, the individual has mechanisms to resist the influence of minorities, although sometimes these have the opposite effect to that intended.

3.2   PRODUCTIVITY OF THE GROUP

The effect that groups have on the performance of an individual are of great interest to our work; knowing if a person performs better or worse by being on a team and the factors that will affect the person are issues that may interest any officer.

3.2.1   GROUP PERFORMANCE

The factors that influence group productivity largely depend on the tasks performed by the group. There are always tasks that are best done alone and others that are best done as a team. To understand this, we will use the taxonomy by Steiner (1972), who divides the tasks into three groups:

1. Divided or unitary tasks: If the tasks can be divided among several people or if they are better suited to individual execution.
2. Tasks of quantity or quality: Work more hours with little productivity or less with greater performance.
3. Relationship between what the individual contributes and what the group contributes to the task: This can be additive (productivity is the sum of the individuals), compensatory (productivity is the average), disjunctive (the group chooses only one person to contribute for the whole group), joint (all contribute), or discretionary (no relation between one person's contribution and the group's contribution).

The importance of this classification lies in the fact that productivity depends on the congruence of the type of task and the way in which the group organizes itself.

We will use Nijstad's (2009) group productivity model with five elements to explain the factors influencing group performance.

- Group members: They contribute skills, motivation, and personality traits. The most important for performance is that the individual and the group goals coincide and the combination of resources is adequate.
- Group tasks: Mainly the type, objectives, etc.
- Processes of group interaction: Internal interaction processes between members and procedures that negatively affect performance must be considered, namely loss of motivation and a decrease in coordination. In the same way, productivity gains can occur due to similar factors.
- Productivity of the group: A distinction is made between the dimensions in which to evaluate productivity (of performance, affective or learning) and performance, as success in the task. The standards can be absolute or normative, relative (comparing with other groups) or individual (comparing with individual performance, if individuals work better alone or in groups).
- The group context: The physical and social environment influences other factors: physical circumstances, time pressure (typical on ships), insufficient resources (again, typical on ships), or the evaluation of others can affect performance.

3.2.2 Effects of the Public and Coercion on Productivity

A key issue when it comes to analyzing the performance of people, is how it can be affected in the presence of others. This question was answered more than 100 years ago when social psychologists discovered that when a subject perceived that he was competing, his performance on physical tasks increased (Allport 1924). Allport already described this phenomenon as *social facilitation*—when a task is executed better just by seeing other subjects doing the same task.

The effect on people witnessing others doing the same task is called the *paradigm of coercion*, while the effect of mere spectators is called the *paradigm of the public*. Likewise, the effects of social inhibition can occur when the task is complex, it includes new situations, or when errors are detected and must be corrected. Social facilitation, therefore, occurs in simple tasks and known situations.

There are several theories to explain these phenomena. The impulse theory of Zajonc (1965) states that we are automatically alert in the presence of others, a state that activates the dominant response in each situation, which is the more likely in any situation. So, if the dominant response is adequate, it will produce facilitation, but otherwise if it is inadequate. This is clearly seen when, for example, an individual runs a marathon and the public and other participants motivate them. Also, when we have to give a speech or deliver a presentation, the first one is a simple physical task and the second is a complex, cognitive case.

Other authors have criticized this theory and have proposed the concepts of *challenge pattern* and *threat pattern* (Blascovich et al. 1999). The challenge pattern occurs when a person has the resources to face the task, otherwise, the threat pattern occurs.

Some authors propose that the effect of distraction can produce an attentional conflict, or the effect of uncertainty when other people are present; therefore, in an evolving situation it is better to maintain a high level of alertness.

However, other authors have conducted experiments showing that the evaluation of others causes a high level of activation, what is called *apprehension by evaluation* (Cottrell et al. 1968).

When spectators participate in the action, this is called the *effect of coercion*. In these cases, people receive information on the performance of others and can compare feedback. This comparison can be of two types: distraction or competition. Distraction can affect productivity; however, competition can help performance, since people are intrinsically prepared to compare themselves with their peers or with slightly better people. This effect of improvement is achieved mainly with easy tasks, but motivation will decrease if the comparison is against people who are less prepared or are notably better prepared.

3.2.2.1 Effects of Coordination and Motivation

The *Ringelmann effect* tells us that the greater the number of people involved in a task, the less effort a subject will make. It may be due to a loss of coordination in a larger group, or to a loss of motivation. These effects are not exclusive, but can occur concurrently; moreover, experimental studies have shown that they can occur simultaneously.

The cause of this effect can be social laziness, as people work less when they think there are others doing the same work, especially when the performance cannot be assessed. This is one of the most consistent effects studied and appears in all cultures. It may be due to three main causes:

1. *Equity in production*: When a person thinks that the other group members are doing the same task, they reduce their performance so that they do not appear to be doing all the work, reducing it in a way that equates to what they think the other members are doing.
2. *Appreciation for the evaluation*: The anonymity of the group makes many people think they are not being evaluated or that they cannot be examined.
3. *Compatibility with the standard*: People can reduce their effort because they do not have a standard with which to adhere.

What most interests officers are the strategies used to avoid social laziness, since as we have seen, it occurs systematically in all human groups:

- Clearly identify the person who is reducing their performance.
- Highlight the individual contributions of each person.
- Indicate the effort that the other group members must make to compensate for the lazy person's deficit.
- Compare performance with other groups (intergroup comparison).
- Highlight the importance of the group's task.

Another phenomenon that occurs is the so-called *problem of the stowaway* or *parasitism*, which happens when a member of the group thinks their work is insignificant or dispensable, a phenomenon that tends to happen in larger groups.

The strategies used to avoid motivational losses are as follows.

- *Social compensation* is an effect that occurs when a person thinks his effort is essential for the group to achieve its objectives, because he believes others do not yield enough, or because they lack motivation (Williams and Karau 1991).
- *The Köhler effect*, which predicts that a person will strive more if he believes the group will fail the task through his fault, due to an increase in motivation for collective well-being (appears greater in the case of women) or competition and status (in the case of men) Kerr, N et al. (2007).

We see the great importance of knowing how an individual performance in a group works to avoid situations of low performance, because one of the main tasks of an officer will be to ensure that his crew performs properly and that he gets the most out of all of them.

3.3   GROUP DECISION PROCESSES: HOW GROUPS MAKE DECISIONS

Following the decision model proposed by Burn (2004), there are six steps to more easily understand how groups make decisions and where problems may occur.

1. Definition of the problem: Objectives of the decision.
2. Identification of alternatives: It is important that there are feasible options and that all are considered, not only those proposed by the leader or pressure groups.
3. Collection of information: Shared by all members.
4. Evaluation of options: Biases that may appear due to personal interests must be controlled.
5. Decision-making: Select an alternative, it can be unanimous or by majority—the stricter the rule, the slower but more satisfactory the process.
6. Implementation: Define the temporal sequence, resources, and the person in charge.

3.3.1 GROUP POLARIZATION

Polarization is one of the most important phenomenon to be considered in group decisions and it is also important if we want to understand how to manage groups and the risks involved in their management.

Polarization is a change in the scale of a decision. Following a discussion, the decision adopted is the dominant one, but more extreme than it was initially. It is not the same as extremity, since extremity is simply the distance from the neutral point of the scale, and polarization changes the whole scale of reference. This phenomenon has received much attention among researchers and is applicable to all types of groups.

Most pertinent to us are the different explanations of the phenomenon, which will help us understand it and ways to avoid or minimize it.

There are two main explanatory processes:

- *Social comparison*: Individuals compare their preferences with others' and then reevaluate their position accordingly. There are two explanations for this phenomenon: *pluralistic ignorance*, in which individuals evaluate the average position of the group and their own, seeking a compromise; and the process of *getting on the winner's car*, here, people foresee what the group's reaction will be, and they give the same answer, but more extreme, improving their position. This effect occurs in all areas, including social and political, in socially desirable decision situations, where the individual will tend to give increasingly polarized responses to what he considers the majority believes, or because it appears in the media or in opinion polls. The mere exposure to the majority opinion is enough to produce polarization.
- *Persuasive argument*: The position of a person depends on the arguments in favor and against which he has. But, if he exposes himself to other new arguments supporting the opinion of the group, polarization will occur. Arguments are persuasive if their validity and their novelty are perceived (they do not have to be real). This phenomenon can also be used to depolarize the group. Empirical studies seem to support this phenomenon more than the social comparison, although both have great repercussions.

We must also look at the theory of self-categorization, which considers that people tend to conform to the norm of the group that is most relevant, in other words, the one that best represents them as a group. This means that people compare themselves with the group prototype and its main characteristics.

It is important to know these processes as they are prevalent in every group, and can sometimes lead to erroneous decisions due to their extremes. As officers, we must be vigilant to these issues when working with groups.

3.3.2 BIASES AND LIMITATIONS IN GROUP DECISION-MAKING

Shared information bias: Groups discuss the information shared by members prior to the discussion more than the non-shared information. This bias may increase if members believe that they do not have enough information to decide, or when it is urgent. This bias is crucial to consider in emergency situations, when insufficient previously shared urgent information can lead members to make wrong decisions. To reduce this bias, exchanging information is essential, but only the most relevant information, because more details imply more bias. All members should also be made aware that their contribution is valuable in decision-making.

Janis (1989) classifies *limitations* into three categories:

- *Cognitive limitations*, when due to the pressure of time or complexity, not all relevant information is considered, leading to insufficient analysis of the problem. This highlights the strategies to satisfy (adopt the first solution that meets minimum requirements) or reaffirms the obvious solution without analyzing possible alternatives.
- *Affiliating limitations* are derived from trying to maintain group harmony, leading to submitting to pressures toward uniformity. This is a dangerous phenomenon that leads group members to try to ingratiate themselves with the leader, ultimately establishing a wall of acquiescence around the leader, where nobody dares to express dissenting opinions. This problem is found in highly syndicated crews, where union leaders can exert pressure of uniformity that affects group performance, by covertly imposing not only their opinion on the group, but also their criteria regarding productivity and the decision-making process, even over the officer in intermediate command. It is therefore a limitation that we must watch for on board ship.
- *Egocentric limitations*, which derive from the desire for control or dominance, or the particular interests of certain group members, leading the group to make biased or incorrect evaluations.

Gouran and Hirokawa (1996) detail some strategies to avoid limitations to group decisions, as follows.

- Avoid applying old solutions to problems, comparing past with current circumstances.
- Find information deficits.
- Motivate to seek the necessary information.

- Try to get a commitment, even if there are disagreements.
- Evaluate alternatives, not people, and do not make personal attacks.
- Be critical of the alternatives.
- Encourage all members to contribute to solutions, thereby avoiding anyone controlling the group.
- Critically evaluate the solutions provided by the members.

3.4   GROUP THINKING

Incorrect decisions made by small groups dominating larger groups led to research on group thinking. The group decision-making model corresponds mainly to Janis and Mann's (1977) model, after several decisions of the US government proved wrong. The model includes three groups of analysis variables: antecedents, symptoms, and defects.

Antecedents: A high level of cohesion is assumed in a decision group. A second group of antecedents is the structural defects of the organization:

- Isolation of the decision group
- Absence of procedural rules for finding solutions
- Insufficient impartial leadership
- Social or ideological homogeneity

Another antecedent group is the *provocative situational context*, which groups external and internal threats, due to either high stress caused by an external risk or to the low self-esteem of the group caused by recent failures.

Symptoms:

- Superiority of the group: The group thinks that it is invulnerable and inherently moral, making failure impossible.
- Mental closure: Negative feedback is rejected and the opposing group is despised in a stereotyped way.
- Pressures toward uniformity: A key symptom on board—and one that must be monitored for its impact on group performance—includes the *illusion of unanimity* and the appearance of the *guardians of the mind*, which prevent the disclosure of information contrary to the prevailing will of the group.
- Defects or failures in decision-making: This may occur due to insufficient consideration of alternatives, inadequate analysis of risks, poor evaluation of alternatives, deficient search for relevant information, and biases in the processing of information.

Finally, and as a practical consequence, Janis proposes some techniques to reduce group thinking:

- Assign all group members as critical evaluators.
- Ensure the impartiality of the leader.
- Subdivide into smaller evaluation and analysis groups.

- In searching for feasible alternatives, gather two or more subgroups to carry out a critical review.
- Discuss the options with trusted people outside the group.
- Invite experts to evaluate decisions before reaching a consensus.
- Rotate the role of devil's advocate. This is a vitally important role in the decision-making process and should not become a simple obligation, but it must make sense, it has to be as meaningful as a vital role in the group.
- Evaluate the threat signals of rivals (if they exist), to control the illusion of invulnerability and prepare contingency plans.
- Before the final agreement, provide an opportunity to express conflicts or doubts.

The conclusions that can be drawn from this study on group polarization are that it is a phenomenon that generally occurs, so we must be prepared to deal with it on board, both in peer-to-peer working groups (among the officers themselves) and with the rest of the crew.

3.4.1 SOME HIGHLIGHTS OF INTERGROUP RELATIONS

It is not our objective to study intergroup relations in depth, but to examine those aspects that may influence conflicts within and between groups.

Among these issues are those that contribute to the emergence of favoritism and rejection. Here, the research suggests that a tendency toward rejection or discrimination against an external group (outgroup) is more consistent than favoritism toward one's own group. Some factors that influence the appearance of these attitudes follow.

- Mummendey and Otten (2001) proved that favoritism occurs when the resources to be distributed are positive; however, favoritism does not occur if negative or aversive tasks are distributed.
- Brewer (1999) also indicates that altruistic behavior is shown to a person's own group, probably because the subject expects reciprocity and it is based on trust. On the contrary, the outgroup provokes greater distrust.
- As we have seen previously, norms, rules, and traditions are factors of group cohesion; therefore, if the individual perceives that the outgroup does not follow these norms, it will be rejected.
- An unstable social context can provoke reactions of rejection toward the outgroup, when competition for limited resources is perceived.
- Media discourses and political ideas also influence the perception of an outgroup.

As guidelines to reduce this rejection of the outgroup, we can focus on reducing intergroup anxiety. The theory of intergroup anxiety (Stephan and Stephan 1985) falls outside our scope of interest, but suffice to say that the mere anticipation of intergroup contact generates anxiety in people. This anxiety depends on several factors, including the frequency of contact and its quality, and not being perceived as contact between equal groups, whether it is voluntary or imposed, competitive or cooperative, etc.

In this sense, it has been proven how contact between members of different groups reduces prejudice, even if indirectly (that is, a member of the same group who knows another group and expresses their view). It is important to foster contact between groups, especially when they are very different, seeking people from dissimilar groups who agree to associate more closely, when it is impossible for big groups to get in touch due to their size.

3.5 BENEFITS OF TEAMWORK

It is not only a matter of reducing individual efforts and taking advantage of the interaction between people, but there are many other benefits derived from working as a team, including:

- Encourages creativity and learning, since the exchange of ideas and points of view engenders individual alternative solutions.
- Mixes talent, because a team is not only a group of different people, but also of different talents, knowledge, and skills.
- Reduces stress, because the workload and responsibilities are shared.
- Improves performance, thus people can concentrate on their strengths, where they perform best, but this can only happen if the group is properly balanced.
- Increases productivity, when the team is weighted and takes advantage of the synergies created between different skills.
- An intelligent business policy is one that promotes the integral development of its members, fostering learning and teaching among its partners. Every organization needs to continuously update their skills and capacities. In this way, members can collectively influence the future development of their organization, having a greater impact than individually.

3.5.1 CHARACTERISTICS OF EFFECTIVE TEAMS

- Participating leadership: The leader must achieve the effective participation of all team members in decision-making, stimulating their allegiance to the group.
- Commitment to the team: The objective of all members must be the achievement of the group's goals, above personal ones.
- Existence of procedures: The responsibilities of each member should be defined according to their personal characteristics and involvement, to avoid conflicts and obtain the best from everyone.
- Criticism and discussion: The team must be open to sound and assertive criticism, as a means to improve its performance.

3.5.2 ROLE OF THE LEADER IN TEAM BUILDING

Although we have already discussed this issue in Chapter 2.1.10 on leadership and intergroup relations, it is important to consider resources that can help in the formation of effective teams. In the merchant navy (and much more in the navy) there has always been an intermediate structure of experienced seamen or non-commissioned officers who have exercised leadership tasks and formed a link between the crew and the officers. Lately, this body has tended to blur or has disappeared completely on many ships. However, experience and theory tells us that these foremen or stewards are of vital importance to the operation of a vessel, and the greater the crew, the more significant they are.

Most of the time these middle managers are given their rank by the shipping company or the ship, but on other occasions the first mate may have the opportunity to decide these positions, and it is here that the skill of the officers is revealed in their analysis of the capabilities and personal characteristics of their crew.

It is also recommended to listen to the crew members and, if possible, let them participate in the decision-making, thereby automatically reinforcing non-imposed leadership. The decision in these cases is not easy, since people with great experience and knowledge may not know how to be good leaders or effectively manage teams. In other cases, it will not be possible, since it will be imposed by the company, so we have to prepare that person to be a leader.

Initially, it is essential to analyze the team context, type of vessel, availability of personnel, and planned rotations (holidays, transfers, etc.). It is also important to know the objectives and the possible temporary duration,it is not the same to form a temporary team for a particular job, than for one that is planned for a longer duration. Some shipping companies opt to only employ permanent crews on their vessels, which greatly facilitates the formation of effective teams. However, other shipping companies impose permanent rotation systems, making it difficult to form teams.

Before starting work on his team, a leader must know the mission, the vision, and the objectives. The mission is the reason for the team, where it is located and what resources it has. Vision is the projection into the future, and should consider difficult but achievable challenges. Finally, the objectives must be clearly and concretely described, so that all members get involved and commit to their achievement.

3.5.3 Stages in Team Building

Although not all teams go through the same stages, the literature describes several phases in the evolution of squads (Table 3.1).

3.5.4 Skills of Team Members

- *Technical or functional*: They are the most basic and every member must master them to achieve their goals.
- *Problem solving and decision-making*: The team must assess the problems they face and develop creative solutions. This process takes place over six phases:
  - Phase I: Recognize the problem and distinguish its causes and effects, for instance, the symptoms of a dysfunction.

- Phase II: Describe the problem, with an analysis and exchange of ideas.
- Phase III: Analyze the cause of the problem, cause–effect diagrams can be used.
- Phase IV: Optional solutions, using the brainstorming technique, then perform a free analysis of the alternatives.
- Phase V: Decision-making—once agreement has been reached on a solution, it is necessary that everyone sticks to it, even if they do not agree.
- Phase VI: Action plan, where chronological development, the people responsible, and the resources are detailed. The action plan requires realistic actions, concrete programs, specific duties, realistic expectations, delegation, and commitment.

3.5.5   SELF-DIRECTED TEAMS

A self-directed team is one that enjoys high autonomy and self-management, including planning and supervising its own activities. This type of team is of concern on board, since it lightens the responsible workload and has a great impact on job satisfaction.

TABLE 3.1

Stages in the Life of the Teams and Role of the Leader

Dependency Stage	Role of the Leader
Uncertainty toward the unknown	Integration, "breaking the ice" among the members
Accurate instructions on tasks are demanded	Identify and share mission and vision
The activities sometimes do not make sense to the members	Identify the objectives
Integration Stage	
Maturity and growth	Generate improvement actions
Conflicts do not threaten the team	Manage conflicts with negotiation
Sense of belonging to the team	Maintain a common purpose
Problems are solved by consensus	Imaginative solutions to problems
Stage of Independence	
Apparent cohesion	Coordinates and directs
Unresolved conflicts remain hidden	Promotes the exchange of ideas; actively listens
Some members hesitate between integrating or preserving their personality	Promotes self-management
Evaluation and role assignment	Select activities aimed at achieving objectives
By becoming aware of the abilities of each member, the group becomes a team	Respect for the members' opinions
Determine the key roles in the team	Encourage participatory decision-making
The team tends toward self-management	He focuses on the results

---

Characteristics to consider are responsibility, planning capacity, communication with the responsible officer, and skills of the members.

The main characteristics of the self-directed teams are

- High performance
- Commitment and efficiency
- Self-correcting actions
- Flexibility and speed
- Promote learning
- Increase individual responsibility
- Promote creativity and innovation

3.5.5.1 Formation of Self-Directed Teams

Figure 3.1 shows the stages in the formation of self-directed teams. To implement actions aimed at improving performance and achieving true self-management, it is necessary to know the skills and capabilities of all members. It is also necessary to know the resources that will be available and those that can be variable depending on the type of vessel, the shipping company, the route, etc. The gradual development of these teams should include self-evaluation and self-regulation, reaching greater levels of autonomy over time.

3.5.5.2 Types of Self-Directed Teams

- Although the scope of these teams will be limited in the nautical environment, it is important to know the contexts in which they can be used. Crews are as varied as the types of ships that exist. Additionally, specific crews are required for ship-building and great shipyard repairs, with external or coupled personnel who, on many occasions, the officer must manage.
- Superior managers are the strategic group as they define the objectives and projects and evaluate the results. We can find these teams in the

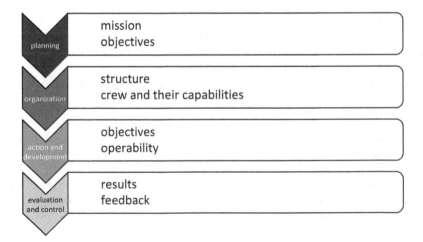

FIGURE 3.1 Stages in the formation of self-directed teams.

ship-building phase, for example, and are composed of engineers, representatives of the shipowner, and the officers themselves.

- Operating groups follow the progress of a project, and analyze and integrate the results. These teams are focused on a specific goal, similar to a special group for firefighting, made up of personnel from different entities.
- Analysis teams are responsible for assessing the results and implementing the necessary improvements. They can be mixed teams from a classification society, for example, where possible deficiencies are evaluated and corrections and improvements are proposed.

The team members' commitments are

- Share the objectives and goals.
- Give positive feedback to other members.
- Communicate problems to higher-ranking members if necessary.
- Report the results.
- Implement or propose improvements to the processes.

3.5.5.3   Role of the Leader in Self-Directed Teams

If the opportunity arises to choose or influence the election of a leader for one of these teams, the following characteristics should be assessed.

- Tendency toward democratic or participative leadership, creativity, assertiveness, and vision of the objectives.
- Technical knowledge above the team average.
- Skills for managing groups—charisma.
- Proactive attitude, organized and disciplined with their superiors.

We must bear in mind that the leader is not only the link between the officer and subordinates, but also represents the team in negotiations and reports their achievements and problems.

3.5.6   Empowerment of Teams

The term *empowerment of teams* refers to granting the decision-making power to the team itself, logically though it also entails the transfer of responsibility. This empowerment will not always be possible, since the group may not meet some essential characteristics for self-management. However, empowerment generates important and valuable influences on group development and productivity, in addition to increasing the sense of dependence and belonging to the organization. All these concepts have an impact on job satisfaction.

3.6   MANAGING DIVERSITY: MULTICULTURAL CREWS

Cultural diversity is one of the topics that has received much attention in social psychology in recent years, with the proliferation of multinational companies, teleworking, and, in general, migratory movements and unstoppable globalization. However, mariners have always worked in multicultural teams, and in a way, working with these teams is part of learning on board ship. However, research on this subject can provide valuable information to improve understanding and cooperation among people from different cultures and increase their personal and work satisfaction and productivity.

3.6.1   DIVERSITY AND ITS EFFECTS ON GROUPS: GENERAL INVESTIGATIONS

Social psychologists seem to agree that diversity produces positive effects not only in terms of flexibility and innovation, but also relationships with other groups. The disadvantages are a higher probability of conflict, a reduction in cohesion, and a greater chance of abandonment (although this in our context is less likely).

One of the leading researches in this field was carried out by Watson, Kumar, and Michaelsen (1993) in a US university, where homogeneous groups were compared with different ethnic groups. At first, the homogeneous groups performed better on the task and had better group processes. However, by the end of the four-month study, the heterogeneous groups had reached the level of the homogeneous groups and had begun to overtake them in aspects such as breadth of perspectives and alternatives generated.

In another study, Audickas, Davis, and Szczepańska (2006) divided groups of Greek (from a more collectivist culture) and Swedish (from a more individualist culture) students into homogeneous and heterogeneous groups, and exposed them to various tasks. They found that homogeneous and individualist groups resolved tasks more quickly, but collectivists had better social interactions, and there were no differences in task effectiveness between the heterogeneous groups. As we can see, several facets must be considered when studying these groups, since some factors intervene in group performance and job satisfaction.

3.6.2   CULTURAL DIMENSIONS

The works of Franke, Hofstede, and Bond (1991) and Hofstede (1980, 1983) are the most cited in the literature on multicultural teams. These authors have identified a series of dimensions that define the culture of a company:

- *Hierarchical distance*: Represents the number of members with less power accepting inequalities in its distribution. For example, countries with a low distance of power are Denmark and New Zealand, while high-distance countries would be Malaysia and Guatemala. Alternatively, although hierarchical distance is negatively associated with individualism, if the level of economic development of the countries under study is controlled, that is, when only nations of low or high economic development are compared, this association disappears (Franke, Hofstede, and Bond 1991). The cultures with high-power distance value the social hierarchy and have respect for authority; this distance is also manifested in the respect for certain professions.

- *Collectivism and individualism*: Individualist cultures promote independence between people and their group, while collectivist cultures value the needs of the group and provide the individual with their identity. Nordic and Anglo-Saxon countries tend to be individualist, while many South America and Asian countries in general are collectivist. South European countries would occupy an intermediate position, but with a tendency toward collectivism. In collectivist societies, individuals offer loyalty to their group in exchange for protection, whereas for individualists loyalty corresponds more to the family and each individual is responsible only to himself.
- *Masculinity and femininity*: According to Franke, Hofstede, and Bond (1991), masculine cultures reinforce differences between the sexes, while typically feminine cultures do not pay attention to gender stereotypes, valuing cooperation and helping the weak. Males are more permissive, they value their quality of life and the complementariness of the sexes. Countries with typically male cultures include Japan, Austria, and Mexico, and countries with typically female cultures include Scandinavia, the Netherlands, Chile, and Costa Rica. Following the same researchers, male cultures are focused on personal achievement and competitiveness, personal success, salary, recognition, and challenge; while the feminine focus is on interpersonal relationships and more harmony.
- *Control of uncertainty*: Tells us about how we face novel, uncertain, or ambiguous situations. Low-control cultures often implement norms to avoid the anxiety produced by novelty; on the contrary, countries that are more flexible and tolerant with the anxiety produced by uncertainty have fewer rules and are more open to changes. In general, Anglo-Saxon countries have a lower score and South American and Asian countries have a higher score.
- *Short-term or long-term orientation*: Cultures with a long-term tendency or orientation value caution, perseverance, and thrift, as well as respect for the elderly; examples are Asian cultures, such as China and Japan. Short-term oriented incentives encourage spending, and quick profits are more important than interpersonal relationships. Western cultures in general are short-term oriented.
- *Complacency or moderation*: Deals with the importance of happiness and the control that each has over his or her life. Cultures with high indices of complacency are more permissive with everything related to the enjoyment of life, while moderates value the suppression of impulses and social norms are stricter. This last dimension is not included in many studies and has recently been added.

For the interested reader, I strongly recommend a visit to the web page of the author of these studies (Hofstede n.d.), where you can make comparisons between different countries in the dimensions of interest, and obtain detailed information on each culture. This visit will be useful when we discuss the management of groups whose characteristics are unknown or we want to delve deeper into their behavior. Figure 3.2 is an example of the graphics that can be obtained on the aforementioned web page.

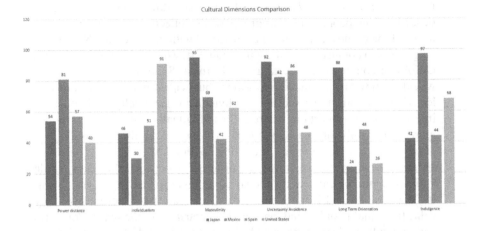

FIGURE 3.2 Comparison of cultural dimensions between four countries. (Data obtained from https://www.hofstede-insights.com/.)

3.6.3 Multiculturalism in the Maritime Context

Fortunately, there is some great research that can guide us in understanding the idiosyncrasies of multicultural teams in the merchant fleet. In this manual, we will focus on summarizing the research and provide suggestions for working with teams.

3.6.3.1 Research on Multiculturalism in the Merchant Navy

The most complete work on this subject is by Horck (2006), where aspects such as language barriers, challenges in teaching seafarers today, and multicultural crews are addressed. Other research is also included, for example, on decision-making in multicultural teams (Horck 2004). The author reflects on the culture in shipping companies, ethnicity (the feeling of belonging to a certain race or ethnic group), and andragogy (the science of teaching adults). He also identifies potential problems that can lead to accidents: insufficient cultural awareness, lack of knowledge of English, and communication obstacles that carry security risks.

Another author of interest is Alfiani (2010) who discusses and reviews the trend of employing multinational/multicultural teams, as well as the sociocultural background and characteristics that distinguish Western and Eastern cultures. The results of research on multicultural crews are examined and analyzed, and literature on the problems that may arise from using mixed teams, as well as the causes of these problems, are discussed.

Another interesting dissertation is by Benton (2005) who focuses on the problems of multicultural crews and puts special emphasis on working with different cultures, which is one of the biggest challenges that sailors face today, with special emphasis on the education of officers. Progoulaki and Roe (2011) and Maria, (n.d.) focus their review on the shipping companies and the socially responsible industry. They provide an extensive review of the literature on cultural problems, with a focus on working and living conditions and crew management.

Finally, the dissertation by Brenker and Worlds (2014) is a summary of the most important aspects of working with multicultural crews, focusing on communication problems. In a large-scale work, Carol-Dekker and Khan (2016) studied the psycho-social stressors that affect crew members of the merchant fleet, in particular, those derived from cultural differences. They concluded that stress; multiculturalism, culture, language, and gender diversity; insufficient interaction with family and friends; loneliness; and fatigue can have a negative impact on the way seafarers face life on board ship, suggesting the implementation of intervention programs with psychologists.

3.6.4 CURRENT PANORAMA OF THE CREWS

In accordance with the Baltic and International Maritime Council (BIMCO) and the International Chamber of Shipping (ICS) (BIMCO and ICS 2015), the number of seafarers worldwide in 2015 was estimated at 1,647,500, of which 774,000 were officers and 873,500 were juniors. The following estimate of global suppliers of seafarers are in order of importance:

Mariners in general: China, Philippines, Indonesia, Russia, and Ukraine
Officers: China, Philippines, India, Indonesia, and Russia
Subalterns: Philippines, China, Indonesia, Russia, and Ukraine

Following the BIMCO report, it was estimated that in 2015 the world merchant fleet comprised 68,723 vessels. The most important category was general cargo vessels making up 31% (mostly container ships) of the total vessels, followed by bulk carriers with 16% and offshore supply vessels with 10%.

The world demand for seafarers in 2015 was estimated at 1,545,000 seafarers, and the industry requires approximately 790,500 officers and 754,500 junior officers. The estimated demand for officers has increased by approximately 24.1% since 2010, while the demand for subordinates has increased by approximately 1.0%. The current situation of supply and demand is a shortage of 16,500 officers and a surplus of 119,000 subordinates, with a surplus of 102,500 seafarers.

In the words of the BIMCO report:

The 2015 report indicates the projected growth in the world merchant fleet over the next ten years, and its anticipated demand for seafarers, will likely continue the trend of a general shortage in the supply of Officers. This is despite the improved levels of recruitment and training and reductions in Officer waste rates in the last five years. (BIMCO and ICS 2015)

This gives us an idea of the current composition of the merchant fleet, where the norm is the coexistence of officers from one or several nations with crews of another nation (unique).

3.6.5 CHALLENGES OF MULTICULTURAL CREWS

3.6.5.1 Living and Working Conditions

Most crews are hired by shipping agents who have little or nothing to do with the conditions of life on board ship and who are usually only concerned with the certification of the crew, so it is easy for a Filipino crew member to remain on board for nine consecutive months, while a crew member from Eastern Europe remains for six months and an officer for only 30–90 days, often depending on the trip. Some union organizations report all kinds of abuses with certain nationalities—extended boarding periods, poor living conditions on board, job instability, and insufficient rest and leisure time.

Gerstenberger (2002) affirms seafarers are working immigrants, but not in the same way as immigrants who move from one country to another, since seafarers do not migrate to another country, but to the world market. The interested reader can dig into the recommended standards and international conventions of life on board on the ILO website (International Labour Organization n.d.).

Most sailors work seven days a week, with long embarkation periods, among them the most prominent are Filipino sailors who endure these long periods away from home. They consider themselves emigrants in a foreign country and make the sacrifice for their family, in this way their perceived health status is good. In general, the state of health perceived by seafarers was good, regardless of their nationality.

The most informative work on the differences between nationalities and the conditions on board is by Jensen et al. (2006), in which 6461 sailors from 11 countries participated.

Obviously, all conditions must be considered when working with crews from different countries. An officer's job is to know the living conditions, the family, and the personal circumstances of his crew and their cultural characteristics such as hierarchical distance, collectivism and so forth.

As stated in Section 3.6.4, most crews come from Eastern cultures, with high collectivism, great distance of power, high control of uncertainty, and little permissiveness; unlike the officers, who may come from Western cultures.

3.6.5.2   Idiomatic Differences

Pyne and Koester (2005) analyzed maritime accidents regarding communication failures, not only between crew members of different countries, but also among native speakers, due to misinterpretations or errors in the reception of the message. Communications are not only problematic when maneuvering, when the ship is at risk, or in navigation on the command bridge, but in day-to-day life on board.

In this sense, there is never enough emphasis on using the IMO (n.d.) Standard Marine Communication Phrases (SMCP); there is a relaxed attitude towards their use on many ships, and an inclination to use more colloquial English that can lead to errors. This is even worse in Anglo-Saxon-speaking crews, who will naturally use their mother tongue instead of the maritime standard. This circumstance is observed daily in vessel traffic services.

Regarding communications in daily life on board ship, in its 1995 revision, the International Convention on Standards of Training, Certification and Watchkeeping for Seafarers (STCW) agreement established English as the common language. However, there are great cultural differences in the interpretation of certain common phrases. Each language uses a series of internal codes that are strongly rooted in the culture itself, thus, language and culture influence and model each other.

For example, to reply "no" in isolation is hard for people from Asian cultures, who are culturally accustomed to answering yes especially to a superior, due to the great power distance typical of their culture. Also, speaking volume causes confusion between different nationalities; for example, Greek people in general tend to speak loudly, which may be seen as offensive to subordinates from Asian cultures. For instance, even among Spanish speakers, Spanish people tend to raise their voices, while South Americans are much more moderate and can interpret a pitch elevation incorrectly. Without doubt, the best way to solve these difficulties is by asking the crew members, encouraging communication outside the guard, and establishing lasting relationships with them, since it is impossible to know in advance all the language and cultural differences.

For many crews, however, their understanding of English is limited, especially in Eastern European and Asian countries, where using the SMCP is imperative in maneuver situations or on the bridge. Similarly, Brenker and Worlds (2014) found in their research that crew members much preferred to interact with others of their own language than with others of a different language, regardless of whether they were officers or subordinates, although among individuals of the same rank interaction is more likely than between different ranks.

It is therefore essential to increase knowledge and use of English for daily interactions, that is, colloquial English, but the SMCP should be used between ships or with shore stations at all times.

Progoulaki and Roe (2011) also report the importance of people's knowledge, finding that many crew members prefer to work with others they already know, even if they are from different cultures, as in the case of Greek and Filipino people in the report. Thus, great importance is given to the maintenance of the crews over time. These ideas match what we already know about group formation and their dynamics.

3.6.6 NON-VERBAL LANGUAGE AND CULTURAL DIFFERENCES

Equally or more important than language is gestural communication or non-verbal communication. It is even said that up to 70% of communication is non-verbal and the detection of emotions plays a key role in relationships within multicultural groups, causing many disagreements and problems in communication and coexistence in such a small space. It is vital to know the cultural differences that exist in various aspects of non-verbal communication to avoid possible conflicts.

This is a much discussed topic within the literature, with abundant research on the subject, especially in recent times, with the increase in globalization. As a reference work, we follow the book by Fort (2005), *Multiculturalism and Communication: Sociocultural Bases for Tourism and Public Relations*. It is an extensive study though obviously not adapted to our subject, but the curious reader will find it of great interest. Following this author, we will discuss various aspects of non-verbal communication that have the most impact on life on board:

- *Paralanguage*: Refers to the way of speaking, the tone, timbre, volume, and emphasis according to the word. The voice conveys spontaneity or professionalism, decisiveness or insecurity, frankness or falsehood, firmness, leadership, etc. We have already commented on the differences that exist in terms of volume, but we are also interested in the differences in terms of tone and pitch. Between Japanese and Scandinavian people, it is normal to control expressivity as a sign of respect, while among Arab and Latino people it is normal to show emotions through exaggeration. Among Nordic people it is normal to respect the turns of words and to vocalize correctly, while Latinos and Mediterranean people usually stop and talk over each other. Asian people tend to be respectful and pause when talking. Arab (especially) and Mediterranean people in general seem to shout when they speak to Nordic or Anglo-Saxon people. Japanese people are restrained to the point that it is difficult for them to ask questions so as not to disturb the other person and sometimes they are unable to respond with a "no" to not be impolite.

- *Haptic communication*: Refers to touch, physical contact itself. In some cultures, physical contact is acceptable, but in others it is totally inappropriate. We will see some of the most common categories on board ship:
  - *Handshake*: Germany: Firm, vigorous, two to three times, plus a faint smile. Arab countries: Soft, repetitive, persistent. France: Soft, fast, and repeated. Asian cultures: Soft and short. Southeast Asia: Soft and persistent. Great Britain: Moderate and short. United States: Firm, energetic, and short, three strokes of medium intensity, stare, and smile. Korea: Moderately firm. Latin countries: Firm and prolonged, touching the forearm or wrist reinforces confidence. Russia: Soft and only once, but many times a day. Europe: Shake hands on arriving and leaving.
  - *Proxemics*: Personal space or *personal bubble*. There are cultures that engage in high physical or tactile contact, such as Latin, Russian, South Asian countries, Arab countries, and French. French people do not usually admit personal contact between strangers, but it is quite frequent between relatives. In Hispanic culture, physical contact between equal classes is allowed. Anglo-Saxons avoid touching, as do Japanese and Nordic people, who demand respect for their personal bubble. The Japanese tend to maintain a larger personal space than the Americans, but the Americans stay farther apart from each other than between Latinos and Arabs. The Anglo-Saxon habit of saying sorry after bumping into someone on the street shows the importance of personal distance to them; in turn, it seems to them that other cultures are intruders in their personal space, and they become distant in the eyes of other cultures. Arab people tend to come so close that they can breathe each other's breath; they tend to seek the truth or the lie through staring, physical closeness, tone of voice, etc. For them, physical proximity is a sign of sincerity, as well as displays of emotion, vehemence, and anger that indicate openness, goodwill, and personal confidence. They even grab the other's hands, shake them, and shout to reach an agreement. Arab people show agreement or friendship by holding each other's hands (it looks like a fight) and pointing at each other with their index finger; this is often observed in the markets when negotiations are established. Mediterranean people come closer to talk and the position is more frontal. Latino and Mediterranean people tend to pay more

attention to who comes closest (offers more availability), but it is the
other way round between Nordic and Asian peoples, and if they stand
back or cross their arms it is better to change tactics.

- *Kinesis*: Body posture. The position of the legs and arms closed or
crossed can show distance or protection. On the contrary, a less rigid
and open arms position generates more confidence, since it demonstrates openness to the ideas of the other. A sitting position can also
facilitate or cool a negotiation, for example, sitting next to someone
shows confidence and closeness, sitting in front demonstrates confrontation or negotiation. The oblique, angled position is better for a chat, as
when two people sit on a sofa. The upright position with arms crossed
denotes confrontation and aggressiveness.

- *Oculesics*: Visual contact. Ocular behavior is essential in humans to
distinguish intention, power, or status, since visual information is the
most important in the processing of information. The dilation of the
pupils is important because when a person shows interest, their eyes
dilate to capture the maximum amount of light. The direct gaze shows
frankness and little fear, but there are cultures where it is used as a sign
of power and superior status over another, which, if it is a subordinate,
must not be maintained. Little or no contact indicates indifference or
distrust. In the United States, you can look at strangers (recognizing
their presence is polite) but do not maintain eye contact as Arab and
Indian people do, or even Latino people. In Asia, eye contact is limited
because it is associated with power and authority, especially Japanese
people, who often avoid looking directly into the other's eyes. British
people look the other way when they talk and look back to indicate
they are finished. In cultures with a long distance of power, keeping
eye contact with a superior is a challenge. Nordic people maintain eye
contact more than Latinos, as an expression of openness and without
fear of being aggressive or dominant because they coexist in contexts of
democratic equality, but they keep their gaze fixed, alternating contact
lapses.

- *Modal communication*: Manners, dress, outward appearance, and taboos
are the aspects covered in this section. Although Western fashion has *de
facto* been imposed, many countries maintain their customs and clothing
and it is important to know their characteristics, to determine status and
power, for example, among Arab people. In Arab, Mediterranean, and Latin
American countries, emotions are openly expressed, whether joy, mourn-
ing, or complaining, while in Asia emotions are controlled, as happens with
friendship or affection. Western cultures have an open facial expression,
but it is more restrained among Asians. The extreme case is Japan, where
it is difficult to recognize emotions. Among Westerners, Mediterranean
people are more expressive than Nordic people in general. Smiling is also
a point to consider, since it denotes frankness and closeness, kindness, and
even subordination. In the United States, smiling shows pleasure, cour-
tesy (forced or not), and dissimulation; people usually smile at strangers in

public (except in big cities). For Russian people it is a mistake to smile and arouses suspicion. In Asia, it serves to control emotions; for example, in Vietnam, it is common to tell a tragedy at the beginning of a story and end with a smile. Westerners are more in control of their laughter and smiling, in Southeast Asia they always laugh, but in Japan they laugh to hide a lapse, error, or disappointment.

- *Chronemics*: Everyone knows that the meaning of time is different depending on the culture, which can lead to serious misunderstandings if one does not know in advance the cultural preferences in this regard. There are two types of cultures according to the temporal sense: monochronic cultures that focus on a single objective and plans are made in advance; and polychronic cultures, where several things are done together and tend more toward improvisation. In polychronic cultures, delays are more acceptable than in monochronic cultures. In Western countries in general, delays are frowned upon, particularly in the case of Anglo-Saxons. However, Arab people see haste as a sign of distrust in a person. In cultures with great distance of power, rushing is frowned upon, as well as getting right to the point of a conversation, and it is advisable to make preambles before addressing a topic.
- *Communication styles*: Styles vary according to the culture, as well as the context in which they are used.
  - *Direct or indirect*: The message can be implicit or explicit. It has much to do with collectivism or individualism, with collectivist cultures more prone to the indirect style. In Asian countries, the indirect style is considered polite and elegant.
  - *Elaborate or succinct*: In Arab countries use an elaborate and rich style of language, using metaphors, etc. In Western countries, there is a tendency to ensure the understanding of the message rather than its form.
  - *Personal or contextual*: These styles are also related to individualism and collectivism. The personal style is informal and shows individualism and equality, while the contextual style depends on status, and the receiver is expected to understand the message more by the context than by the message itself.
  - *Instrumental or affective*: Instrumental or affective also fits with individualism and low context, and collectivism and high context. The instrumental is oriented to the objectives and goals of which it speaks, but the affective is oriented to the relation and to the receiver. Nowadays, the instrumental type is mostly used in negotiations.
- *Cultural rules of speech*: Each culture has its own rules; for example, Nordic people speak little but think a lot about what they are going to say, while Americans give great importance to everyone talking, even if they do not contribute anything interesting, which can make them seem superficial. The turn to speak is usually agreed with gestures or a look, and in some countries (Mediterranean, for example) interruptions are tolerated, but the Germans and Anglo-Saxons see it as bad taste.

- *Gesture*: Certain gestures are interpreted differently depending on the cultural environment. In general, people from Nordic cultures are less expressive than Southern Mediterranean and Latin people.

3.6.6.1   Ambiguous Gestures and Cultural Differences

Various gestures can be misinterpreted, so when in doubt, the best action is to keep your hands quiet. The main cultural differences in gestures are

The left hand: It is impure for Muslims, Hindus, and Buddhists. One must avoid using it to touch objects and people or deliver documents. The variety of gestures that differ between cultures is wide, so we will only focus on the main ones considering the variety of crews of the international fleet.

Showing the sole of the shoe: This is an insult for Arab and Thai peoples, as it is the dirtiest part of the body. Do not cross your legs showing the soles of the shoes, just above the knee and without lifting the sole.

In most non-Western countries, women should sit with their knees together, otherwise it is interpreted as a sexual overture. Hitting the fist of one hand on the palm of the other has an obscene meaning in Asia.

The raised thumb or index finger has an obscene meaning in many countries. Do not use the index finger to call anyone, e.g. waiters, for this, it is better to extend the palm and move the fingers.

In France, Italy, and Germany, touching the temple with the index finger when looking at someone indicates that the person is stupid. However, in Spain and the United Kingdom, it means "I'm thinking." In the Netherlands, a raised vertical index finger is telling you that you are smart, but if it is horizontal, they are calling you an idiot. In Japan, touching the temple with the index finger turned forward means "to be thinking," and if it is turned backwards it indicates madness. Both this and others depend on other non-verbal signs, but it is better to avoid them in order to not create confusion.

Thumb raised: Another gesture to avoid. In many countries it means "great," in European countries such as Germany and Italy, it indicates the number 1, but in other Mediterranean and Arab countries it has sexual connotations. It is obscene in Nigeria, and can be in Australia.

"V" for victory: This gesture is done by showing the palm of the hand, but if the back of the hand is shown it is a serious insult in the United Kingdom.

The OK sign: Another gesture to avoid. The thumb and index finger in a circle means "all goes well" in many cultures, but for the Japanese it indicates "coins" or "we are going to talk about money"; in France it means "0" or "it is not worth anything"; in Russia it is a sexual innuendo; and in Latin America and Turkey it indicates that a man is homosexual.

Whistle: Its meaning in the United States is approval, but in most countries it is the opposite.

Raising eyebrows: In Germany this indicates admiration; for British people it indicates skepticism; and for Arab people, denial.

Eye winking is considered obscene in Australia and Taiwan.

In Japan, to indicate someone is lying, they suck the index finger and pass it across an eyebrow; head scratching means dislike; and to show disagreement or rejection, they move the hand back and forth as if to separate the idea or proposal.

In Greece, showing and extending the palm of your hand to someone is called *moutza* and is a serious insult. Beware of using this gesture among a Greek crew.

Shaking hands is an international greeting, but not so much in Asia. If we interact with Asian crew members it is likely that they are used to shaking hands, but at the beginning it is better to wait for them to initiate the greeting, which may be the "namaste" in India and its surroundings or the "saikerei" (Japan) or "wai" in Thailand, tilting the head or body.

Greeting women can be complex. In Western crews shaking hands is usual, but in almost all Asian and Muslim cultures it is better to wait, rather than to cause an uncomfortable situation.

In many Arab countries, stroking the mustache is a direct sexual response to women. It is better to avoid this gesture in front of women.

3.6.6.2   Some Negotiation Rules

Although most of the negotiation rules that appear in the literature focus on the business world, many can be used on board ship when working with a crew from different countries. There are some tactics that can make things easier.

There are two types of negotiation style, *competitive* and *cooperative*, to reach a compromise or an agreement. The distance of power has an important role to play and it is essential to understand this aspect as we choose a style to follow in negotiations.

Thus, for example, before Asian people get to the point of a negotiation, they have to go through several preliminaries to gain the trust of the other. Anglo-Saxons may interpret a delay in a decision as a trick or weapon, rather than as a pause to think. Japanese people and Anglo-Saxons seek agreement rather than confrontation, but Mediterranean people are more prone to bargaining or negotiating approaches. In Russia and Eastern countries, nepotism and family favors are common and are considered acceptable, so it is usual for these crew members to allow family members on board ship as personal favors and it is usually highly valued.

In collectivist cultures the harmony of the group is highly valued, while in individualists it gets more to the point. For Anglo-Saxons, reciprocity agreements or small favors are normal, but for Russians it is considered a sign of weakness. For Arab people, bargaining is a way of life and is by no means seen badly, so one can expect flattery and other typical bargaining gimmicks. For cultures of great distance of power, it is convenient to maintain that distance so as not to lose status, strength, and respect; however, in Western cultures this can be perceived as authoritarianism.

3.6.6.3   Considerations of Religious Differences

An essential part of a culture is religion, and in many countries, it permeates the whole life of the people. For this reason, it is essential to understand the behavior characteristics and the limitations that a certain religion can impose on a crew, out of respect for their traditions.

It is well known that during the month of Ramadan, Muslims abstain from eating and drinking during the day, which can be a serious limitation in a shift-working system, where stopping work can be unacceptable to the shipowner.

Not all Muslims have the same level of compliance with religious precepts, and depending on their origin, some rest can be negotiated in exchange for a certain flexibility. This will be different depending on the crew's origins or where the ship is geographically, because it is not the same in Morocco, Saudi Arabia, or Indonesia.

Islam, besides influencing the religious sphere of the individual, affects their way of acting, relating, focusing on business, etc. The strict application of Islam does not differentiate between religion and state, unlike in the West, where there has long been a separation between church and state. In addition, we must bear in mind that sometimes there are problems of coexistence with other religions, especially Hindus and Jews, which we must take into consideration when establishing multicultural teams.

In general, for Westerners, religion does not impose itself on other aspects of life on board ship, but we must consider that each crew member has a vision of life more or less marked by their religion.

For Russians and Ukrainians, for example, their Orthodox beliefs are important, as well as for Catholic Poles. In general, the more to the West, the more secularized society is; therefore, religion enters more into the personal sphere and has much less influence on public aspects of life and work. Considering the distribution of crews by country, religions found on board will be mostly Catholic (Filipinos and South Americans), Buddhist (Chinese and Southeast Asians), and Orthodox (Greeks, Russians, and Ukrainians).

Readers interested in deepening their knowledge of these aspects has a good reference in the book by Levey and Modood (2009), *Secularism, Religion and Multicultural Citizenship*.

3.6.7 GENDER DIFFERENCES ON BOARD

According to Carol-Dekker and Khan (2016), discrimination against women on board can make the work environment stressful for women, reaching extreme cases such as the South African Akhona Geveza, who committed suicide after being raped by an officer. The subsequent investigation revealed that abusive practices against female cadets were common in the fleet. In an extensive paper, Manzano (2014) reported that sexual harassment is not only aimed at female seafarers, but some male navigators become vulnerable when they do not fit into the male hegemonic category.

During an interview, a Filipino male sailor shared his feelings of being snubbed by seafarers, stating that he was reserved by nature and avoided conflict with his crew mates. His demure nature led him to be labeled "gay" and he was socially isolated, despite not being homosexual.

The attitude toward women in many cultures (especially Asian cultures) can range from paternalism to more or less veiled harassment, as many men do not approve of women on board, especially in certain positions that may require physical strength. This can lead to condescension and to considering woman inferior.

Although in Western countries it is common to see female officers, there is still a long way to go before reaching full equality. To promote equality, the IMO (n.d.)

has launched a series of films focused on women working on ships. In textual words, "more women are needed on board, especially in leadership roles." In addition, the IMO program on the integration of women in the maritime sector (integrated workplace management system [IWMS]) has as its main objective to encourage the member states to open the doors of their maritime institutes to allow women to train with men, acquiring the highest level of competence. For the interested reader, the World Maritime University has compiled a selection of articles on this subject in the book *Maritime Women: Global Leadership* (Kitada, Williams, and Froholdt 2015). The IMO policy is directed by the Busan Declaration (IMO 2013), where the organization is committed to fighting for the effective integration of women on an equal footing with men in the maritime sector, literally, "work to raise awareness of the role of women as a valuable resource for the maritime industry."

On the website Marine Insight (Mukherjee 2017), a Philippine female officer posted her insights on the challenges a woman faces on board, including:

- The problem of being accepted on board, of earning the respect of men and having to constantly prove their worth.
- The fight against loneliness, which can be overcome with hobbies, study, and doing pleasant activities.
- Managing prejudices and stereotypes, especially in certain societies, where women are condemned to remain in the home. Resilience is a virtue in these cases. Unfortunately, many men still insist on traditional roles for women. The only advice that can be given here is resist and follow your dream without letting anything or anyone distract you.
- The culture on board is focused on men, and this is something that will take time to change, for instance, equipment, uniforms, etc. We must encourage the shipping companies to adapt the gear and general ergonomics to the requirements of women.
- Face Cognitive Closure. If you have a good relationship with a partner, this can arouse suspicion and jealousy in others, which can raise misgivings of favoritism. In these cases, it is useful to try connecting with everyone equally.
- Adapt to physical and mental fatigue, because for a woman some physical jobs can be exhausting, in addition to the constant need to prove their worth. Many women may feel that they need to "overwork" to show they can tolerate difficult tasks, which might be a mistake.
- Prove your competence, as we have seen, is one of the main mental and physical challenges facing a woman on board; and it is likely that this challenge of constantly leaving one's comfort zone is a stimulus to overcome and explore what one is capable of, that is, to face the heavier tasks as a personal challenge, rather than as a problem.
- Gather courage to do difficult tasks as the physical and mental challenges on board can sometimes be overwhelming. Life on board and a sailor's job is not easy even for experienced men, so resilience is a personality trait that women on board must cultivate.

3.7 WHEN ALL TEAMWORK FAILS: THE CASE OF THE *BOW MARINER*

The chemical tanker *Bow Mariner*, sailing under the Singaporean flag, caught fire, exploded, and finally sank off the coast of Virginia (USA) on February 28, 2004. The crew was cleaning dangerous residuals (methyl tert-butyl ether [MTBE]), when some explosions took place. Of the 27 crew members, 18 were reported missing and 1 was recovered dead. The US Coast Guard (USCG) carried out an investigation, and the reader can download the full report from the USCG web page:

https://www.dco.uscg.mil/Portals/9/DCO/Documents/5p/CG-5PC/INV/docs/ documents/bowmar1.pdf.

All certificates pertaining to the vessel and the crew were correct at the time of the accident, and the lifesaving and firefighting equipment were also in proper condition. The vessel was also equipped with an inert gas system.

The ship's voyage began in Al Jubail, Saudi Arabia, and after bunkering in Algeciras (Spain), departed for New York (USA). Crossing the Atlantic, the *Bow Mariner* encountered heavy weather, so the vessel was delayed for two days more than expected. The vessel discharged part of its cargo in New York and departed for Houston, Texas, on February 27. At the time of the accident the wind was from NW 15 knots with 2 m waves, the water temperature was only 5.5°C, and the air temperature was 4.4°C. The tanker was carrying 12,070,584 L of ethyl alcohol on 10 centered tanks, and the remaining 22 tanks contained only MTBE residuals.

3.7.1 THE CREW

The crew was composed of Greek officers and Filipino seamen. The captain was 52 years old and finally disappeared into the sea.

No alcohol or drugs were detected in survivors' blood. The routines on board appeared to comply with STCW rules to prevent fatigue; however, survivors reported that the chief mate's working routine in port was exhausting (which is very common in tankers). The investigation found some non-compliance with ISM code, mainly regarding procedures.

3.7.2 THE EVENTS

At 10:00, while on watch the third officer (who was Filipino) was told by the captain to open the 22 tanks and proceed with the cleaning. Those tanks were full of MTBE vapors, and as the cleaning process was taking place, there was a strong smell of alcohol in the air. Some security procedures were broken to prevent ignition, for instance, two crew members were blowing down cargo lines with compressed air, probably causing static electricity. At 18:00, the deck lights were switched on, and soon the first explosion was heard and the ship started to list to starboard.

Then, the catastrophe. The Greek officers grouped together and began to discuss between themselves in Greek, issuing no instructions to the crew. The lifeboats were blocked due to the ship's listing. The captain didn't order a distress call, so the third officer had to act by himself. The crew started to leave the vessel in total disarray and confusion, trying to reach some life rafts. Three people reached a life raft and tried desperately to find survivors. Eventually, they found two survivors covered in fuel, but alive. At 19:37, the *Bow Mariner* sank.

3.7.3   THE HUMAN FACTOR

The ISM code that was not fully implemented related to the human element:

- Fire and boat drills never took place (most probably).
- Training scheduled only when inspection due.
- Second assistant engineer didn't have a proper takeover and was on his first voyage with the company. In fact, it was a *two-men crossing at the gangway* takeover.
- There were no familiarization procedures, arguing lack of time.
- Administrative procedures were unknown to the crew.
- There were no established procedures for cleaning tanks, the crew only followed instructions, without questioning them.

In addition, the existing company culture gave absolute authority to the captain, forcing the crew to obey orders without question. On board, the differences between the Greek and Filipino officers were remarkable. The Filipino officers did not take their meals in the officers' mess, lacked responsibility, and were closely supervised. Orders were given verbally and not always in a polite manner.

The official report states that lack of trust was normal practice in the company, and the chief officer performed almost all the important duties on board, resulting in a lack of preparation of the Filipino crew members. There was also an atmosphere of distrust and a lack of confidence. There were cases when the Greek officers verbally abused the Filipinos, and these attitudes were confirmed when inspections took place on other vessels owned by the same company.

Evidence of the lack of a team culture on board was clear during the emergency, when Greek officers spoke in Greek, although the official language on board was English. The situation was chaotic and led to the disaster. Other officers tried to explain that the captain's behavior was due to emotional trauma caused by the explosions and listing, but as we know, it is essential to be familiar with drills in those circumstances, and they never took place.

While a breach of safety procedures led to the accident, we must focus on the human behavior and group practices, on which the USCG report is very clear. The captain failed in his duties: lacking proper and necessary leadership, refusing to organize the crew, and giving up on his most elementary duties.

The captain established an authoritarian style of leadership that formed two differentiated teams with no connection between them. He kept vital information from the majority of the crew, giving them no opportunity to even express their opinions.

Cultural differentiation also played a role in the overall climate on board. Disrespect, distrust, and information concealment were usual practices, while the

Filipinos feared losing their jobs if they verbalized their opinions. This total lack of teamwork and toxic organizational climate played a vital role in the disaster and loss of life.

3.8 CONCLUSIONS

In general, teams found on board arrive already formed and will usually have a rigid and closed structure, so it will be more important to understand the cultural and social differences that we are going to manage, any possible sources of conflict concerning other cultures that may exist on board, and the policy of the shipping company regarding relay management, transshipments, etc. The officer's mission is to understand his crew members, to know them thoroughly in their context, and to know how to anticipate possible sources of conflict.

REFERENCES

Alfiani, Didin Susetyo. 2010. "Multinational and Multicultural Seafarers and MET Students: A Socio-Cultural Study for Improving Maritime Safety and the Education of Seafarers." *World Maritime University Dissertations.* 425.

Allport, Floyd H. 1924. "The Group Fallacy in Relation to Social Science." *American Journal of Sociology* 29 (6): 688–706.

Aritzeta, Aitor, Sabino Ayestaran, and Stephen Swailes. 2005. "Team Role Preference and Conflict Management Styles." *International Journal of Conflict Management* 16 (2): 157–182.

Asch, Solomon E. 1956. "Studies of Independence and Conformity: I. A Minority of One against a Unanimous Majority." *Psychological Monographs: General and Applied* 70 (9): 1–70.

Asch, Solomon E, and H Guetzkow. 1951. "Effects of Group Pressure upon the Modification and Distortion of Judgments." In Harold S Guetzkow (ed.) *Groups, Leadership, and Men* (pp. 222–236). Carnegie Press.

Audickas, Simonas, Charles Davis, and Magda Szczepańska. 2006. "Effects of Group Cultural Differences on Task Performance and Socialization Behaviours." *Europe's Journal of Psychology* 2 (1).

Benne, Kenneth D, and Paul Sheats. 1948. "Functional Roles of Group Members." *Journal of Social Issues* 4 (2): 41–49.

Benton, G. 2005. "Multicultural Crews and the Culture of Globalization." In *International Association of Maritime Universities (IAMU) 6th Annual General Assembly and Conference.* WIT Press.

BIMCO, and ICS. 2015. "Man Power Report 2015."

Blascovich, Jim, Wendy Berry Mendes, Sarah B Hunter, and Kristen Salomon. 1999. "Social 'Facilitation' as Challenge and Threat." *Journal of Personality and Social Psychology* 77 (1): 68–77.

Bond, Rod, and Peter B Smith. 1996. "Culture and Conformity: A Meta-Analysis of Studies Using Asch's (1952b, 1956) Line Judgment Task." *Psychological Bulletin* 119 (1): 111–137.

Brenker, Michael, and Complex Working Worlds. 2014. "Teamwork across Cultures: Communication in Multinational Seafaring Crews Teamwork across Cultures: Agenda," No. July.

Brewer, Marilynn B. 1999. "The Psychology of Prejudice: Ingroup Love and Outgroup Hate?" *Journal of Social Issues* 55 (3): 429–444.

Burn, Shawn Meghan. 2004. *Groups: Theory and Practice*. Wadsworth. Recording for the blind and dyslexic.

Carol-Dekker, Lydia, and Sultan Sultan. 2016. "Reflections on the Psycho-Social Distress within the International Merchant Navy Seafaring Community." *Journal of Psychology* 7 (2): 53–60. http://krepublishers.com/02-Journals/JP/JP-07-0-000-16-Web/JP-07-2-16-Abst-PDF/JP-07-2-053-16-182-Khan-S/JP-07-2-053-16-182-Khan-S-Tx[1].pdf.

Coleman, Janet Fagan, Robert R Blake, and Jane Srygley Mouton. 1958. "Task Difficulty and Conformity Pressures." *The Journal of Abnormal and Social Psychology* 57 (1): 120–122.

Cottrell, Nickolas B, Dennis L Wack, Gary J Sekerak, and Robert H Rittle. 1968. "Social Facilitation of Dominant Responses by the Presence of an Audience and the Mere Presence of Others." *Journal of Personality and Social Psychology* 9 (3): 245–250.

Crutchfield, Richard S. 1955. "Conformity and Character." *American Psychologist* 10 (5): 191–198.

Deutsch, Morton, and Harold B Gerard. 1955. "A Study of Normative and Informational Social Influences upon Individual Judgment." *The Journal of Abnormal and Social Psychology* 51 (3): 629–636.

Eagly, Alice H, and Linda L Carli. 1981. "Sex of Researchers and Sex-Typed Communications as Determinants of Sex Differences in Influenceability: A Meta-Analysis of Social Influence Studies." *Psychological Bulletin* 90 (1): 1–20.

Festinger, Leon. 1950. "Informal Social Communication." *Psychological Review* 57 (5): 271–282.

Fort, C A. 2005. *Multiculturalidad y Comunicación: Bases Socioculturales Para Turismo y Relaciones Públicas*. Laertes. https://books.google.es/books?id=Tg6sAAAACAAJ.

Franke, Richard H, Geert Hofstede, and Michael H Bond. 1991. "Cultural Roots of Economic Performance: A Research Note." *Strategic Management Journal* 12 (S1): 165–173.

Gerstenberger, Heide. 2002. "Cost Elements with a Soul." In *Proceedings of the 9th International Association of Maritime Economists Conference*. Panama, November 13–15, 2002.

Gouran, Dennis S, and Randy Y Hirokawa. 1996. "Functional Theory and Communication in Decision-Making and Problem-Solving Groups: An Expanded View." In R Y Hirokawa and M S Poole (eds) *Communication and Group Decision Making* (pp. 55–80). Sage.

Hochbaum, Godfrey M. 1954. "The Relation between Group Members' Self-Confidence and Their Reactions to Group Pressures to Uniformity." *American Sociological Review* 19 (6): 678–687.

Hofstede, Geert. 2018 "Home—Hofstede Insights." Accessed May 14, 2018. https://www.hofstede-insights.com/.

Hofstede, Geert. 1980. "Motivation, Leadership, and Organization: Do American Theories Apply Abroad?" *Organizational Dynamics* 9 (1): 42–63.

Hofstede, Geert. 1983. "The Cultural Relativity of Organizational Practices and Theories." *Journal of International Business Studies* 14 (2): 75–89.

Horck, Jan. 2004. "An Analysis of Decision-Making Processes in Multicultural Maritime Scenarios." *Maritime Policy and Management* 31 (1): 15–29.

Horck, Jan. 2006. *Jan Horck a Mixed Crew Complement*. Malmö University.

IMO. n.d. "Standard Marine Communication Phrases." Accessed May 15, 2018a. http://www.imo.org/en/OurWork/Safety/Navigation/Pages/StandardMarineCommunicationPhrases.aspx.

IMO. 2018. "Women in the Maritime Industry." Accessed May 19, 2018b. http://www.imo.org/en/MediaCentre/HotTopics/women/Pages/default.aspx.

IMO. 2013. *Busan Declaration*. IMO Regional Conference on the Development of a Global Strategy for Women Seafarers, Busan, South Korea, April 16–19, 2013.

Insko, Chester A, Sarah Drenan, Michael R Solomon, Richard Smith, and Terry J Wade. 1983. "Conformity as a Function of the Consistency of Positive Self-Evaluation with Being Liked and Being Right." *Journal of Experimental Social Psychology* 19 (4): 341–358.

International Labour Organization. 2018 "International Labour Standards on Seafarers." Accessed May 14, 2018. http://www.ilo.org/global/standards/subjects-covered-by-international-labour-standards/seafarers/lang--en/index.htm.

Janis, Irving L, and Leon Mann. 1977. *Decision Making: A Psychological Analysis of Conflict, Choice, and Commitment.* Free Press.

Janis, Irving Lester. 1989. *Crucial Decisions: Leadership in Policymaking and Crisis Management.* Simon & Schuster.

Jensen, Olaf C, Jens F L Sørensen, Michelle Thomas, M Luisa Canals, Nebojsa Nikolic, and Yunping Hu. 2006. "Working Conditions in International Seafaring." *Occupational Medicine* 56 (6): 393–397.

Jones, Edward Ellsworth, and Harold Gerard. 1967. *Foundations of Social Psychology.* Wiley.

Kerr, Norbert L., et al. "Psychological mechanisms underlying the Köhler motivation gain." *Personality and Social Psychology Bulletin* 33.6 (2007): 828–841.

Kitada, Momoko, Erin Williams, and Lisa Loloma Froholdt. 2015. *Maritime Women: Global Leadership.* Springer.

Levey, Geoffrey Brahm, and Tariq Modood. 2009. *Secularism, Religion and Multicultural Citizenship.* Cambridge University Press.

Lott, Albert J, and Bernice E Lott. 1961. "Group Cohesiveness, Communication Level, and Conformity." *Journal of Abnormal and Social Psychology* 62 (2): 408–412.

Maass, Anne, and Russell D Clark. 1984. "Hidden Impact of Minorities: Fifteen Years of Minority Influence Research." *Psychological Bulletin* 95 (3): 428–450.

Manzano, Joanne V. 2014. "Filipino Crosscurrents: Oceanographies of Seafaring, Masculinities, and Globalization [Book Review]." *Philippine Studies: Historical and Ethnographic Viewpoints* 62 (2): 289–292.

Maria, Progoulaki. 2018 "Management of Cultural Diversity: Identifying the Training Needs of Seafarers and Shore-Based Personnel in the European Maritime Shipping Industry." Paper delivered at the 20th International Conference on "Managing Complexity in Shipping and Port Markets: Firms' Business Models, Co-operative Games and Innovative Public–Private Interactions," International Association of Maritime Economists, July 3–5, Marseille, France.

Marques, José M, and Dario Paez. 1994. "The 'Black Sheep Effect': Social Categorization, Rejection of Ingroup Deviates, and Perception of Group Variability." *European Review of Social Psychology* 5 (1): 37–68.

Moreland, Richard L, and John M Levine. 1992. "The Composition of Small Groups." In B Markovsky, C Ridgeway, and H Walker (eds) *Advances in Group Processes*, Vol. 9 (pp. 237–280). JAI Press.

Moscovici, Serge. 1981. "On Social Representations." In J P Forgas (ed.) *Social Cognition: Perspectives on Everyday Understanding* (pp. 181–209). Academic Press.

Mukherjee, Paromita. 2017. "A Woman Seafarer Describes the Challenges She Faces on Board Ships." https://www.marineinsight.com/life-at-sea/woman-seafarer-ship-challenges/.

Mummendey, Amélie, and Sabine Otten. 2001. "Aversive Discrimination." In Rupert Brown and Samuel L Gaertner (eds) *Blackwell Handbook of Social Psychology: Intergroup Processes* (pp. 112–132). Blackwell.

Nijstad, Bernard A. 2009. *Group Performance.* Psychology Press.

Progoulaki, Maria, and Michael Roe. 2011. "Dealing with Multicultural Human Resources in a Socially Responsible Manner: A Focus on the Maritime Industry." *WMU Journal of Maritime Affairs* 10 (1): 7–23.

Pyne, Robyn, and Thomas Koester. 2005. "Methods and Means for Analysis of Crew Communication in the Maritime Domain." *The Archives of Transport* 17 (3–4): 193–208.

Sherif, Muzafer. 1936. *The Psychology of Social Norms.* Harper & Brothers.

Steiner, I D. 1972. *Group Processes and Group Productivity.* Academic Press.

Stephan, Walter G, and Cookie White Stephan. 1985. "Intergroup Anxiety." *Journal of Social Issues* 41 (3): 157–175.

Terry, Deborah J, and Michael A Hogg. 1996. "Group Norms and the Attitude–Behavior Relationship: A Role for Group Identification." *Personality and Social Psychology Bulletin* 22 (8): 776–793.

Watson, Warren E, Kamalesh Kumar, and Larry K Michaelsen. 1993. "Cultural Diversity's Impact on Interaction Process and Performance: Comparing Homogeneous and Diverse Task Groups." *Academy of Management Journal* 36 (3): 590–602.

Williams, Kipling D, and Steven J Karau. 1991. "Social Loafing and Social Compensation: The Effects of Expectations of Co-Worker Performance." *Journal of Personality and Social Psychology* 61 (4): 570–581.

Zajonc, Robert B. 1965. "Social Facilitation." *Science* 149 (3681): 269–274.

4 The Psychological Challenges of Life on Board

What are the characteristics of life on board ship and how do they influence people's behavior? Can they be improved during the sailor's time at sea, making it more comfortable for sailors? What behaviors influence personal performance and productivity? We will attempt to provide some answers for these and other questions. We have previously focused on the behavior of human groups, but there is significant room for improving personal life on board, understanding the psychosocial challenges of a sailor's life, and adopting adaptive behaviors to the special work environment.

In recent years, awareness of the challenges of life on board has increased. The works of Jensen et al. (2004, 2006) are significant due to their size. A total of 64,161 sailors participated in the study, and among some of its conclusions it is noteworthy that seafarers from some countries with long periods at sea had a better self-assessment of their health than those who had shorter periods. The duration of time at sea as a variable was not a sufficient indicator to predict the health status of seafarers nor were differences found in their state of health according to nationality.

Oldenburg et al. (2009) specify the circumstances that may cause stress and other health problems on board, noting that a seafarer's work cannot be compared to any other work ashore.

In particular, Oldenburg et al. complain that there is insufficient current field scientific studies on the work and life situation of ship crews. They also say that "the crew of a multi-ethnic ship can aggravate the feeling of isolation of seafarers and can increase the psycho social problems associated with the long separation from the family." In addition, they detect a growing concern about fatigue and cognitive overload. In another work by Oldenburg, Hogan, and Jensen (2013), the authors review field studies carried out to date, focused on fatigue, where they observe that tiredness does not seem to depend on the age of the seafarer and is often associated with poor sleep quality, noise, and night shifts. They conclude that more field studies are needed on stressors other than fatigue. Fieldwork carried out by Oldenburg et al. (2009) on a German fleet, identified both physical and psychological stressors reported by seafarers; these are shown in Figure 4.1.

Using the current bibliography available and reviewing various books and articles, Carol-Dekker and Khan (2016) reflect on the psychosocial factors that affect seafarers. They conclude that stress/distress, multiculturalism, culture, language, gender diversity, insufficient interaction with family and friends, loneliness, and fatigue can have a negative impact on life on board ship.

PERCENTAGE OF RESPONSES
(% OF PEOPLE WHO CONSIDER IT A PROBLEM)

FIGURE 4.1   Physical and psychosocial stressors identified on German vessels. (Data from Oldenburg et al. 2009.)

Pauksztat (2017b, 2017a) discussed fatigue and stressors on board, pointing out that social support is important to maintain health. He also studied the wearying effects of the ship's route, the stops in port, the difficulty of tasks, and the physical demands as sources of stress and tiredness.

Regarding circadian cycles (sleep–wake) and their effect on seafarers, Reyner and Baulk (1998) conducted a pilot study with crews of ferries, and detected a great disturbance in sleep patterns, with loss of cognitive abilities. In this sense, we can highlight the *Martha Project* (Warsash Maritime Academy n.d.), which includes fatigue prevention software and has studied the effect of fatigue on captains, their performance and cultural differences in crews (InterManager and Warsash Maritime Academy 2016). The effect of fatigue on maritime accidents has been addressed by Xhelilaj and Lapa (2010) among others. The authors support the idea that fatigue has a detrimental effect on seafarers, which in turn can impact on the normal operation of ships at sea and thus can be implicated in accidents. Finally, Arendt et al. (2006) studied sleep and circadian rhythms, monitoring several crew members on an Atlantic voyage. Among other findings, they concluded that crew members undergoing rotating shifts had worse sleep quality.

We will analyze the psychosocial stressors that affect seafarers and offer some guidelines to cope with them. The order in which they appear is not significant, because for each person, that will be a relative factor.

4.1 SHIFT WORK: THE CIRCADIAN RHYTHMS

The operation of ships, normally 24/7, except ferries and other vessels in coastal traffic, generates work schedules based on shifts, which cause a mismatch between working hours and rest, thereby disturbing the circadian rhythms. These circumstances are relative and not the same for each person; there are those who can and cannot adapt favorably to shift work.

In addition, there are two main types of shifts or sea guards: four- and six-hour shifts. Usually these shifts are fixed by the shipping company or the regulations in each country. Accidents related to fatigue occur more often in six-hour shifts than in four-hour shifts. In a sample of 185 bridge officers who worked a system of two shifts per day, Härmä et al. (2008) found that officers reported shorter and more micro-sleep dreams during the guards compared to officers on the three-shifts system (7.3% vs 1.5%). Obviously having eight hours of rest between shifts provides a more restful sleep than having only six hours.

Evidence indicates that shift work negatively influences workers' physiology, health, safety, and sleep and circadian cycles, resulting in altered sleep during the day and excessive sleepiness during work hours, leading to so-called shift work disorder (SWD) (Kalmbach et al. 2015).

Recently, these authors affirmed that working shifts has consequences for depression and anxiety. Åkerstedt and Wright (2009) also found that the quality of cognitive processes is optimal when the internal clock thinks it is daylight, so the problem depends not only on the night shift but also on the cycles of change, which prevent the body from properly adjusting its biological clock. They also found "a greater boost to homeostatic sleep results in impaired cognition, increased sleepiness and increased sleep propensity". Drowsiness and fatigue are problems frequently reported in several investigations (Akerstedt 1998; Åkerstedt and Wright 2009; Shwetha et al. (2012), n.d.; Härmä et al. 2002; Saksvik et al. 2011; Wang et al. 2011).

Why are circadian rhythms so important? Circadian rhythms are a series of physical, mental, and behavioral cyclical changes that occur daily and are guided by lighting patterns. These rhythms are present in all living beings, both plants and animals, since the day/night cycle affects all living beings. These rhythms are coordinated by the so-called biological clock, which consists of a group of approximately 20,000 neurons that form a structure called the "suprachiasmatic nucleus" (NSQ). The NSQ is located in a part of the brain called the hypothalamus and receives direct information from the eyes, because sunlight activates or deactivates it.

4.1.1 SLEEP AND ITS DISORDERS

Circadian rhythms influence sleep–wake cycles, hormonal secretions, dietary habits and digestion, body temperature, and other important bodily functions. An unbalanced biological clock can produce altered or abnormal circadian rhythms.

Irregular rhythms have been linked to several chronic medical conditions, such as sleep disorders, obesity, diabetes, depression, bipolar disorder, and seasonal affective disorder. Thus, the sleep–wake cycles have a physical and psychological effect on people.

The circadian rhythms adjust sleep through the production of melatonin. When sunlight decreases, the NSQ emits a signal to the body to start making melatonin, and this causes sleep. In the shift-working system, the body continuously fights nature that dictates when and when not to sleep.

According to Nogareda Cuixart and Nogareda Cuixart (1995), alterations in eating habits can occur, because people usually eat fast meals at unusual hours, normally with a high caloric content, coupled with a large consumption of coffee or other caffeinated beverages.

However, the most serious changes occur during sleep. There are two main sleep phases: a light nap and a deeper sleep. It is during this deep phase that muscle tone decreases and the physical recovery of the body occurs. During the first phase, paradoxical sleep occurs when psychic recovery takes place, dreams are produced, and metabolic and endocrine activity increases.

These phases occur cyclically during the rest periods, about four times each. In shift work, for example, paradoxical sleep is reduced in a morning shift, because one has to wake up too early and shorten the last hours of sleep. A reduction in deep sleep is observed in the night shift, due to the alteration in sleep/wakefulness, and due to the greater difficulty sleeping during the day (daylight, everyday noises, etc.), making recovery from physical fatigue difficult. All these circumstances lead to a state of chronic fatigue that can produce physical symptoms including headache, irritability, depression, various digestive diseases, and circulatory system problems (Nogareda Cuixart and Nogareda Cuixart 1995).

Insufficient sleep can also lead to perceptual alterations, impaired judgment, insufficient concentration, memory failure, etc. The MARTHA Project (InterManager and Warsash Maritime Academy 2016) found that officers in general are more tired at work and suffer from poorer quality sleep and a higher level of stress than their subalterns. These findings are in line with other studies. Discrepancies between European and Chinese seafarers were also found, indicating some interesting differences between the two dissimilar management-style cultures, with Chinese crews having poorer sleep quality, but lower stress levels than European crews.

The MARTHA Project also asked about preferred type of sleep. Some people prefer to sleep at night and some prefer to do so in the morning. Curiously, and contrary to the general population, the majority of the sailors identified themselves as evening types, which could be due to the effort made by all sailors to adapt their sleep to the shifts system. It is not surprising that it is the second officer (who normally enters guard at four in the morning) who has the worst quality of sleep. Also, as factors that indicate drowsiness: new regulations to take care of, duties and requirements for seafarers, insufficient proper maintenance, work in port, increased inspections and paperwork, bad conditions of the cabins, working on a new ship, and the quality and professionalism of teammates.

How can problems derived from sleep disturbance be avoided or reduced? The following are guidelines to make the shift system more tolerable:

- Take naps and eat well. A short nap (about 30 minutes) improves performance and subjective feelings of well-being (e.g., it may be useful to take a nap in the afternoon, before the night shift). Rest rooms, dining rooms, etc., should be offered on board, so that the night shift workers can rest and eat a balanced meal, even during the shift if possible.
- Install adequate lighting in the workspace. Exposure to sunlight is an effective treatment for sleep disorders. An intense light in the evening/night delays the circadian rhythms (allowing a person to be active at night and aiding sleep during the day). Special daylight lamps (about 3000 lx) can be used during night work to activate the body. Sunlight should be avoided before going to sleep (e.g., wearing dark glasses, otherwise the light will inhibit melatonin) and the room should be in complete darkness. Use daylight lamps. In order to stay awake and alert, lighting that simulates daylight should be used. Therefore, the bridge and the machine room should use this type of lighting, and monitors, data presentation screens, etc., should be adjusted to keep the operator alert. At night, the bridge remains in darkness; however, a chart room or similar room should use this type of light.
- Use melatonin supplements. As we have seen, melatonin has a fundamental function in sleep, and is currently found in many preparations that, together with medicinal plants, can help regulate sleep. In their reference work on melatonin suppression due to shift work, Burch et al. (2005) found serious alterations in the production of melatonin in these workers. Thus, melatonin can be recommended as a sleep inducer, although with due caution. In general, field studies show its effectiveness, but it is variable and depends on the person. Its long-term effects are still not clear and it seems to adversely affects people with Type 2 diabetes. In general, specialists agree that it should be taken between one and three hours before going to sleep and it should not be used continuously as a precaution. Basically, melatonin should not be used for more than two weeks, and during that period you should try to adapt the organism naturally. Also, if one experiences side effects such as headaches or excessive sleepiness, intake of melatonin should be curtailed.
- Establish a regular schedule for sleeping and eating, which seems a basic measure, but is difficult to carry out on board ship.
- Use caffeine to stay awake. Probably the only drug that can be recommended, since many studies have demonstrated the impact that caffeine has on the level of alertness and the low dependence it generates (Burch et al. 2005; Muehlbach and Walsh 1995; Schweitzer et al. 2006; Walsh, Muehlbach, and Schweitzer 1995). Caffeine should not be consumed at least three hours before going to sleep, as its effects on the body can last and can completely disturb sleep. It should also be noted that American-type coffee contains more caffeine, 100 mg per cup compared to 30 or 40 mg in an espresso. It is also a slow absorption drug, so the first noticeable effect of drinking coffee is actually placebo.
- Set up a sleep hygiene routine. Sleep hygiene techniques are simple guidelines and tips to help improve rest quality. Some may be difficult to

implement on board ship, because following certain routines can be complicated when you have to get up two hours after going to bed for a maneuver at two in the morning, for example.

- Only light food should be consumed before sleep, and never go to sleep feeling hungry. Hot milk has been proven to aid sleep. Sugary foods should not be consumed because sugar promotes alertness.
- Live an active life, exercise regularly, adapted to your age and abilities.
- Avoid long naps, never more than 30 minutes.
- Maintain a regular sleep schedule.
- Avoid mental activity in bed, e.g., reading, watching TV, or using a computer; particularly TVs and computers as they emit blue light tricking the brain into thinking it is daytime. Do not make *overnight thinking*, try not to bring problems to bed. A bed should be associated with sleep and should not to be used as a meditation space because an inadequate reinforcement association could be established. Any problems should be written down and considered in daylight with a clear mind.
- Create an adequate environment that fosters sleep, e.g., temperature, light, and noise. This is not always possible on board ship. The entire crew must be encouraged to respect other crew members' rest.
- Have a ritual before going to sleep. Behavioral psychology teaches us that if we associate certain behaviors with sleep, they will reinforce the sleep response. Brushing teeth, putting on pyjamas, showering in the same way and in the same order, can be reinforcing behaviors.
- If you are not sleepy, get up. If you have not been able to go to sleep for more than half an hour, it's time to get up and try to relax. For this purpose, techniques of paradoxical intention can be used, that is, achieving the opposite of what is intended, for example, trying not to fall asleep watching an interesting TV program or reading a good book—the more you try to stay awake, paradoxically the more drowsy you become. It is an effect that we have all experienced and it can be useful in these cases.
- Avoid using alcohol as an aid to falling asleep. Although alcohol may promote the onset of sleep, it is also associated with earlier awakenings, sleep disturbances, and poor sleep quality. Drinking too regularly increases the risk of long-term damage to your physical and mental health, your work, and your social and personal relationships. In addition, alcohol can nullify rapid eye movement (REM) sleep or the light sleep phase, inhibiting a restorative sleep.

4.1.2 Fatigue

The International Maritime Organization (IMO) describes fatigue as a "state of tiredness or drowsiness that results from prolonged mental or physical work, exposure to a hostile environment or loss of sleep that may impair performance and reduce alertness" (IMO 2001, p. 1).

In addition, the IMO's circular MSC/Circ.813/MEPC/Circ.330 describes the consequences of fatigue on humans as a "reduction in physical and mental capacity due to physical, mental or emotional exertion that may damage almost all physical abilities, including: strength, speed, reaction time, coordination, decision-making, or balance" (IMO 2001).

Fortunately, there is much research on the effects of fatigue on board ship, e.g., a Google Scholar search returns more than 60,000 results. To cite some of the most relevant, Xhelilaj and Lapa (2010) claimed that fatigue is the main cause of maritime accidents, in addition to contributing to a diminished work performance, which appears as a guideline in all research on the subject. Undoubtedly, the most interesting research for us is the program Seafarer Fatigue: The Cardiff Research Program (Smith, Allen, and Wadsworth 2006), carried out by the University of Cardiff, and all the associated research (Reyner and Baulk 1998; Smith 2007). The interested reader can read this publication, as our focus is on the effects of fatigue and some key strategies to combat it.

4.1.2.1 Factors That Contribute to Fatigue

The IMO distinguishes several types of factors that contribute to chronic or pathological fatigue. The following is a summary of several of the factors, emphasizing the main ones and adding a few more.

Specific factors pertaining to the crew: Sleep, circadian cycles, and rest periods. Stress, monotony, and boredom. Health, diet, and disease. Personal problems and relationships with colleagues. Insufficient skill or expertise in the position and insufficient training. Alcohol or drugs use. Age and workload.

- Organization: Personnel policies, transfers, and rotations. Relations with ground crew. Paperwork. Low salaries and various work problems, especially excessive overtime. Training of the crew (by excess or defect). Linked to the trip, including frequency and duration of port calls, type of route, weather, density of traffic and guards in port, which can be more exhausting than those of the sea.
- Factors of the ship: Ship's design, ergonomics in general, level of automation, and redundancy of tasks. Equipment operation, hardware faults. Maintenance and age of the vessel. Movement of the ship, for example, roll-on/off are usually horrendous to sleep (among other types). Accommodation spaces and their design and maintenance. Food (this factor is ignored by the IMO; however, experience indicates that the importance of a good cook and a responsible shipping company in terms of food are paramount).
- Environmental factors: Temperature, inside and outside the accommodation, especially in the engine room. Humidity. Excessive noise levels. Movement of the ship both for work and for rest. Route, traffic density, and so forth.

4.1.2.2 Effects of Fatigue

Fatigue should be considered a physical and psychological condition. It is also subjective in terms of perception, citing the definition of the University of Cardiff (1996,

p. 5): "Fatigue is a consequence of continuously high levels of information or prolonged work overload it implies subjective feelings of fatigue or insufficient inclination to work." In other words, a broader view of this definition describes acute fatigue or chronic fatigue, like depression, drowsiness, stress, disturbed circadian rhythm, or boredom.

Obviously, as recognized by the IMO, the working environment of the seafarer marks the great difference with other jobs, since the seafarer cannot leave his job when he finishes his day, he is a captive worker, with all the psychological implications that this entails. The main effects of fatigue on a sailor are

- Decrease in alert level. A fatigued person is not only sleepy, but also will not respond with the same speed in risky situations. They are unable to maintain situational awareness while on the bridge, that is, they are unable to process information and make decisions with the required fluidity.
- Decrease in performance level, especially memory lapses, delays in making decisions, or failures of decisions made. Physically, fatigue can lead one to be apathetic in the workplace, lacking initiative, or not collaborating.
- Communication problems, especially if a non-native language is being used.
- Decrease in the level of vigilance, with the possibility of micro-sleep occurring, which further increases the feeling of tiredness. Of primary importance in situations of low visibility.

Table 4.1 summarizes the symptoms of fatigue that the IMO includes in its report, suitably adapted.

4.1.2.3   Strategies to Combat Fatigue On Board

- Plan the workload, especially paperwork. Reduce it as much as possible during the night watch, especially between 03:00 and 06:00. During the night, attention must be exclusively on navigation.
- Be active and have a healthy lifestyle, avoid excess weight and a sedentary lifestyle. Walk around the deck, or use the gym if available. Take walks on land when in port. Socialize. Monitor health and report any problems to a doctor as soon as possible.
- Avoid a ship's maneuvers as much as possible during the critical hours of dawn.
- Get up and walk regularly during the watch. Exit to the bridge wing if you feel drowsy (and it is feasible).
- Shipping companies could consider the implementation of automated anti-fatigue warning systems, *dead man* systems, and similar that exist in the market (Optalert n.d.).
- Consider the ergonomics of the workplace, which should be comfortable and well ventilated.
- Shipping companies should contemplate the three-shift guard system as the best one for the seafarer.
- If you are prone to dizziness, a proven natural remedy is ginger pills.

TABLE 4.1
Symptoms of Fatigue in Officers

Effects	Signs and Symptoms
Decrease in Performance	Unable to organize a series of activities. Worried about a single task and focuses on a trivial problem, neglecting the most important ones. Reverts to old but ineffective habits. Less vigilant than usual.
Inability to concentrate	Judges distance, speed, time, etc. incorrectly
Decreased ability to make decisions	He does not appreciate the seriousness of the situation. Ignores situations that should be monitored. Chooses risky options. Difficulty with simple arithmetic, geometry, etc.
Poor memory	He does not remember the sequence of elements of a task. Difficulty remembering events or procedures. He forgets to complete a task or part of a task.
Slow response	Responds slowly (if at all) to normal, abnormal, and emergency situations.
Control of body movements	He may seem drunk. Inability to stay awake. The speech is affected, but he can crawl, his movements can be slower or distorted. Sensation of heaviness in arms and legs. Decreased ability to lift, push, or pull. Greater frequency of dropping objects such as tools or parts.
Humor changes	Calmer, less talkative than usual. Unusually irritable. Greater intolerance and antisocial behavior. Depression.
Changes in attitude	He does not anticipate the danger and does not observe or obey warning signs. He seems unaware of his own deficient performance. Too willing to take risks. Ignores normal controls and procedures. Shows an attitude of "I do not care." Weakness in handling things. Distaste for work.

Source: Adapted from IMO (2001, p. 20).

- Posing novel and rewarding tasks, studying to improve work, playing an instrument, photography, painting, etc.
- Schedule drills for the crew so that the personnel resting are not disturbed.
- Keep the crew structure already established and with proven optimum performance.
- A greater involvement of the shipping companies in the promotion of a healthy lifestyle for their crews is essential, not only for social responsibility, but as a productive investment. Do not skimp on training, and seek early and preventive medical attention for every crew member.
- Schedule in advance the tasks of each crew member while the ship is in port, ensuring that every crew member can enjoy their stay on land.
- Remember the international guidelines and agreements on fatigue, stress, etc. These should be posted in the office of every captain and first officer:
  - International Labor Organization (ILO): Convention on the hours of work and manning of seafarers' ships—ILO Convention No. 180 ("International Labour Standards on Seafarers" n.d.). Of particular interest are the annexes.

- International Maritime Organization (IMO): International Convention on Standards of Training, Certification, and Watchkeeping for Seafarers, 1978, as amended in 1995 (STCW Convention) (IMO 2010); Seafarers' Training, Certification, and Guard Code (IMO 2010) Parties A and B; International Security Management Code (ISM Code) (IMO-OMI n.d.); and several other guidelines/recommendations.

4.1.3 UPROOTING

Uprooting is defined as the stress caused by the separation of a sailor from his family and friends. Seafaring is a hard profession, and involves being permanently on the job with long periods away from family. The study by Oldenburg and Jensen (2012) revealed that 70.2% of non-European seafarers and 54.0% of European seafarers considered that separation from family due to long periods of work on board ship was stressful.

Many seafarers fear that these periods of separation from their families will lead to a progressive loss of contact and social isolation from society in their own country, all aggravated by the fact that crews are increasingly reduced, which limits the social contact on board ship.

Feelings of separation and loneliness can lead to depressive situations, triggering psychological fatigue and work stress, and feelings of being less involved in the work. Unfortunately, each person reacts differently to this problem and the only protective factor is resilience.

Age also plays an important role, because when you have small children, family separation is much more difficult.

Traditionally, many sailors have fought loneliness using alcohol and tobacco, which cannot in any way be recommended. A way to guard against loneliness is through maintaining communication with family and friends, made easier by satellite communications and mobile phones. Keeping yourself occupied and having a good relationship with your colleagues on board ship is also good advice.

4.1.4 LONG PERIODS ON BOARD SHIP

One of the main stressors identified by Oldenburg et al. (2009) is long stays on board ship. Although there are great differences in terms of nationalities and periods of stay on board, it is possible that on the same ship there are officers with periods of 30 days and subordinates with up to 12 months in a row. This stressor is strongly related to family separation and the feelings of loneliness. The only viable solution is in the hands of shipping companies, by making shipping periods more flexible as a means not only of responsible business management, but also to maintain a motivated crew and encourage a better performance.

4.1.5 MULTI-ETHNIC CREWS

As we have seen in Section 3.6.4 (current panorama of the crews), most crews tend to be multinational and multi-ethnic and this can be a source of conflict.

Responsible management can avoid such sources of conflict, hence the importance that this manual attaches to this topic. It is essential that cultural differences between crew members are known in order to avoid conflicts, and any that arise should be quickly resolved. We must also consider the relative weight of the different nationalities on board ship, in case a certain group becomes a source of influence that reduces the hierarchical authority of the vessel. This issue becomes more critical as crew sizes increase; the rise of the cruise ships fleet means more and more ships have large crews.

The best advice that can be given to overcome the problems derived from multinational crews is to have knowledge of and empathy for different cultures, and to demonstrate respect for the different cultures.

4.1.6 WORKLOAD AND TIME PRESSURE

The work of a sailor has always been marked by workload and time pressure, two factors that are aggravated by paperwork, which many shipping companies seem to have increased despite the digital revolution and paperless office. Officers are also subject to the pressure of decision-making. The only alternative to this problem is the implementation (now possible with telecommunication technologies) of a management system focused on downloading bureaucratic tasks that can be carried out by on-shore offices or resolved by computer systems. This will only be possible if the officers needs are heard and met, since the main mission of a command is not to fill out forms, but the safe and proper functioning of the ship and its crew.

4.1.7 DEMANDS OF THE CHARGE

Finally, within the stressors identified by Oldenburg et al. (2009), some are inherent to the position, especially for the captain, including insufficient qualification of the crew, responsibility for the work of other crew members, and conflicts between security and economic demands.

On these stressors the only possible solution is the alluded responsible personnel policy from the shipping company and a good officer preparation to exercise leadership and create crews that are truly committed to the personnel on board and understand it as a team.

4.2 FATIGUE AND NATURAL TRAGEDY: THE CASE OF THE *EXXON VALDEZ*

Probably the most famous maritime disaster in terms of oil pollution and environmental damage was that caused by the ultra large crude carrier (ULCC) *Exxon Valdez* and it is also the most salient case of a disaster caused by fatigue.

The *Exxon Valdez* was carrying 1,264,155 barrels of crude oil from Alyeska Marine Terminal in Alaska bound for Long Beach (California) when the tanker struck the rocks in Prince William Sound, Alaska, on March 24, 1989, and spilled 35,000 mt of crude oil resulting in an ecological catastrophe. The remote location of the spill hindered the efforts to clean the water and save the sea life. The oil covered over 2,100 km of coastline and 28,000 km² of ocean.

Long-term and short-term effects of the oil spill have been reported. The more immediate effects were the deaths of 250,000 seabirds, at least 2,800 sea otters, approximately 12 river otters, 300 harbor seals, 247 bald eagles, and 22 orcs, and an unknown number of salmon and herring. The effects of the spill could still be seen as late as 2014, when some studies reported the existence of oil on rocks and sands. The economic costs for Exxon were at least $507.5 million in punitive damages and other costs.

The official report can be downloaded from https://www.ntsb.gov/investigations/AccidentReports/Reports/MAR9004.pdf.

4.2.1   THE CREW

The master, who was 42 years old, received a degree in maritime transportation from New York University in 1968. He served on nine tankers with no problems, and he was alternate master of the *Exxon Valdez* previously. He had made no less than 100 passes through Prince William Sound. The *Exxon Valdez* was also considered the best vessel in the company's fleet.

The officer on watch (OOW) at the time of the accident was the third officer, who had received his license in 1986, joined the vessel in February 1989, and had made no less than 18 passages through Prince William Sound. The company records of this individual were excellent, and his assessments were sometimes above normal.

As on many tankers, the crew management could have been improved. The crew did not break the sea watches when loading/unloading, but the chief officer was in charge of the operations at all times. On occasion, the second and third officer had to work extra hours to assist the chief officer and allow him some rest.

The helmsman had served as an able seaman (AB) since 1965, and had previously worked on 19 vessels, including the *Exxon Valdez*. The problem was that he was not an official helmsman, but only an AB. His performance of tasks on board was above average.

The crew consisted of master, first, second and third deck officers, chief, first, second and third engineers, one radio officer, three AB, and three maintenance crews, The crew also comprised a steward, cook, and so on, making a total of 19 crew members.

The coast guard statutes lay down that there must be six hours of rest before taking watch on deck operations. According to the US Coast Guard (USCG), the manning levels on board the *Exxon Valdez* were typical of most tankers (as we know, all over the fleet).

4.2.2   THE EVENTS

The *Exxon Valdez*' equipment was tested and working properly before departure. The pilot proceeded with the required caution to avoid ice and shallow waters. The master decided to leave the traffic lane in order to avoid the ice, and contacted the Vessel Traffic Services (VTS) for permission to do so. Everything appeared to be going as usual; however, the third officer opted not to slow down and proceeded ahead avoiding the ice. The OOW failed to keep track of the vessel's location. The captain had left him alone, when the company's procedures state that he or the chief officer must be on the bridge in shallow waters. The third officer then made some bad decisions based on incorrect positioning and slow and easy reactions to the vessel's course. Finally, after looking at the radar, he realized that there was a serious situation and ordered hard to starboard. He subsequently telephoned the master to inform him that the accident could not be avoided. Attempts made to free the vessel probably caused more oil to spill.

The investigation found that the master's decision to leave the traffic separation scheme to avoid ice was safe, but required two officers on the bridge. Instead, the vessel was only operated by the third officer, who was not licensed for this task alone in those waters. Also, the VTS failed to locate the ship and that she was running into danger.

4.2.3 THE HUMAN FACTOR

The investigation report is very clear in stating the causes of this accident: the main reason was the fatigue of the OOW, who was unable to get a full resting period before leaving the terminal; he had a stressful working day when loading the ship. He was relieving the chief mate of his duties to give him some rest, avoiding having his own necessary time to repose. The third officer would have had only four sleep hours before sea watch, and only a two-hour nap in the afternoon. It is very likely that he had only five to six hours of sleep before the accident (probably less).

The demanding activities didn't stop when the vessel was sailing, because the situation between narrow waters and the presence of ice, were very stressful, given his experience and knowledge. The usual practice on tankers is to have the chief mate in charge of all charging/discharging operations, and impose a six hours on six hours off watch on the other officers. In some tankers, the master takes over part of the navigation duties to relieve his crew of excessive work, but this wasn't the case. This disaster could have been avoided if the vessel had remained in the terminal for the hours needed to provide the crew with enough rest.

Also, there were some failures on the part of the coast guard monitoring the working conditions as they failed to detect that safety was compromised because the vessel was operating with a reduced crew. There was an insufficient number of crew members in the deck department to properly cover the watch in congested waters. There was evidence that there was insufficient crew to carry out maintenance duties and watches at the helm. The reduced manning practices of the Exxon were considered unsafe in terms of the risk of accidents. Moreover, other company practices rewarded overwork despite the safety issue.

There was some evidence of the master's drinking activities before leaving Valdez (by his own admission in court, he drank "two or three vodkas" between 4:30 and 6:30 that same night). Also, the Exxon supervisor was unaware that the master had an alcohol dependency problem and the safety board concluded that he had to be removed from the fleet until that problem was solved. However, during the trial the master was cleared of being drunk during the accident, which is an ongoing controversy.

The VTS also failed as not only was it unaware of the accident, but also it had not tracked the tanker as it sailed through a traffic separation scheme in the presence of ice. Exxon, as well, failed to properly maintain the Raytheon Collision Avoidance System (RAYCAS) radar, which would have alerted the third officer to the grounding.

This catastrophe was a catalog of human errors and wrong management practices that exemplify the importance of the human element on board ship.

REFERENCES

Akerstedt, Torbjörn. 1998. "Shift Work and Disturbed Sleep/Wakefulness". *Sleep Medicine Reviews* 2 (2): 117–28.

Åkerstedt, Torbjörn, and Kenneth P Wright. 2009. "Sleep Loss and Fatigue in Shift Work and Shift Work Disorder". *Sleep Medicine Clinics* 4 (2): 257–71.

Arendt, Josephine, Benita Middleton, Peter Williams, Gavin Francis, and Claire Luke. 2006. "Sleep and Circadian Phase in a Ship's Crew". *Journal of Biological Rhythms* 21 (3): 214–21.

Burch, James B, Michael G Yost, Wendy Johnson, and Emily Allen. 2005. "Melatonin, Sleep, and Shift Work Adaptation." *Journal of Occupational and Environmental Medicine* 47 (9): 893–901.

Carol-Dekker, Lydia, and Sultan Sultan. 2016. "Reflections on the Psycho-Social Distress within the International Merchant Navy Seafaring Community". *Journal of Psychology* 7 (2): 53–60. http://krepublishers.com/02-Journals/JP/JP-07-0-000-16-Web/JP-07-2-16-Abst-PDF/JP-07-2-053-16-182-Khan-S/JP-07-2-053-16-182-Khan-S-Tx[1].pdf.

Costa, Giovanni. 1996. "The Impact of Shift and Night Work on Health." *Applied Ergonomics* 27 (1): 9–16.

Härmä, Mikko, Markku Partinen, Risto Repo, Matti Sorsa, and Pertti Siivonen. 2008. "Effects of 6/6 and 4/8 Watch Systems on Sleepiness among Bridge Officers". *Chronobiology International* 25 (2–3): 413–423.

Härmä, Miko, Mikael Sallinen, Radu Ranta, Tuomas P Mutanen, and K Müller. 2002. "The Effect of an Irregular Shift System on Sleepiness at Work in Train Drivers and Railway Traffic Controllers". *Journal of Sleep Research* 11 (2): 141–151.

IMO. 2001. "Guidance on Fatigue Mitigation and Management". *MSC/Circulation* 1014: 83.

STCW, IMO. "Standards Training." *Certification and Watchkeeping*, IMO Publication, London(2010).

IMO-OMI. n.d. "ISM Code". Accessed May 23, 2018. http://www.imo.org/en/OurWork/HumanElement/SafetyManagement/Pages/ISMCode.aspx.

InterManager, and Warsash Maritime Academy. 2016. "Project MARTHA the Final Report. http://ftp.elabor8.co.uk/martha/flipbook/martha/thefinalreport/mobile/index.html.

"International Labour Standards on Seafarers". 2018 Accessed May 14, 2018. http://www.ilo.org/global/standards/subjects-covered-by-international-labour-standards/seafarers/lang--en/index.htm.

Jensen, Olaf C, Jens F L Sørensen, Linda Kaerlev, M LuisaCanals, Nebojsa Nikolic, and Heikki Saarni. 2004. "Self-Reported Injuries among Seafarers. Questionnaire validity and results from an international study". *Accident Analysis and Prevention* 36 (3): 405–13.

Jensen, Olaf C, Jens F L Sørensen, Michelle Thomas, M Luisa Canals, Nebojsa Nikolic, and Yunping Hu. 2006. "Working Conditions in International Seafaring". *Occupational Medicine* 56 (6): 393–97.

Kalmbach, David A, Vivek Pillai, Philip Cheng, J Todd Arnedt, and Christopher L Drake. 2015. "Shift Work Disorder, Depression, and Anxiety in the Transition to Rotating Shifts: The Role of Sleep Reactivity". *Sleep Medicine* 16 (12): 1532–1538.

Muehlbach, Mark J, and James K Walsh. 1995. "The Effects of Caffeine on Simulated Night-Shift Work and Subsequent Daytime Sleep". *Sleep* 18 (1): 22–29.

Nogareda Cuixart, Clotilde, and Silvia Nogareda Cuixart. 1995. "Trabajo a Turnos y Nocturno: Aspectos Organizativos". Instituto Nacional de Seguridad e Higiene En El Trabajo, 1–10. http://www.insht.es/InshtWeb/Contenidos/Documentacion/FichasTecnicas/NTP/Ficheros/401a500/ntp_455.pdf.

Oldenburg, Marcus, B Hogan, and Hans Joachim Jensen. 2013. "Systematic Review of Maritime Field Studies about Stress and Strain in Seafaring". *International Archives of Occupational and Environmental Health* 86 (1): 1–15.

Oldenburg, Marcus, and Hans Joachim Jensen. 2012. "Merchant Seafaring: A Changing and Hazardous Occupation". *Occupational and Environmental Medicine* 69 (9): 685–88.

Oldenburg, Marcus, Hans Joachim Jensen, Ute Latza, and Xaver Baur. 2009. "Seafaring Stressors Aboard Merchant and Passenger Ships". *International Journal of Public Health* 54 (2): 96–105.

Optalert. n.d. "Control de la Fatiga". Accessed May 23, 2018. http://www.optalert.com/spanish/fatigue-management.

Pauksztat, Birgit. 2017a. "Effects of Job Demands and Social Interactions on Fatigue in Short Sea Cargo Shipping". *Maritime Policy and Management* 44 (5): 623–40.

Pauksztat, Birgit. 2017b. "'Only Work and Sleep': Seafarers' Perceptions of Job Demands of Short Sea Cargo Shipping Lines and Their Effects on Work and Life on Board". *Maritime Policy and Management* 44 (7): 899–915.

Reyner, Louise, and Stuart Baulk. 1998. *Fatigue in Ferry Crews: A Pilot Study.* Seafarers International Research Centre, 1–4. http://www.sirc.cf.ac.uk/uploads/publications/Fatigue in Ferry Crews.pdf.

Saksvik, Ingvild B, Bjørn Bjorvatn, Hilde Hetland, Gro M Sandal, and Ståle Pallesen. 2011. "Individual Differences in Tolerance to Shift Work - A Systematic Review". *Sleep Medicine Reviews* 15 (4): 221–35.

Schweitzer, Paula K, Angela C Randazzo, Kara Stone, Milton Erman, and James K Walsh. 2006. "Laboratory and Field Studies of Naps and Caffeine as Practical Countermeasures for Sleep-Wake Problems Associated with Night Work". *Sleep* 29 (1): 39–50.

Shwetha, Bijavara, and Honnamachanahalli Sudhakar. "Influence of shift work on cognitive performance in male business process outsourcing employees." *Indian Journal of Cccupational and Environmental Medicine* 16.3 (2012): 114.

Smith, Andrew Paul. 2007. "Adequate Crewing and Seafarers' Fatigue: The International Perspective". Centre for Occupational and Health Psychology, Cardiff University.

Smith, Andy, Paul Allen, and Emma Wadsworth. 2006. "Seafarer Fatigue: The Cardiff Research Programme," no. November.

University of Wales, Cardiff. 1996. "Research Workshop on Fatigue in the Maritime Industry (1996: Seafarers International Research Centre) & Seafarers International Research Centre (1996). Proceedings of a Research Workshop on Fatigue in the Maritime Industry, Seafarers International Research Centre."

Walsh, James K, Mark J Muehlbach, and Paula K Schweitzer. 1995. "Hypnotics and Caffeine as Countermeasures for Shiftwork-Related Sleepiness and Sleep Disturbance". *Journal of Sleep Research* 4 (S2): 80–83.

Wang, Xin Shelley, Miranda E. G. Armstrong, Benjamin J. Cairns, Timothy J. Key, and Ruth C. Travis. 2011. "Shift Work and Chronic Disease: The Epidemiological Evidence". *Occupational Medicine* 61 (6): 443–4; author reply 444. https://doi.org/10.1093/occmed/kqr078.

Warsash Maritime Academy. n.d. "Horizon Project - MARTHA, Research Projects at Warsash Maritime Academy". Accessed May 20, 2018. https://www.warsashacademy.co.uk/about/our-expertise/maritime-research-centre/horizon-project/martha.aspx.

Xhelilaj, Ermal, and Kristofor Lapa. 2010. "The Role of Human Fatigue Factor Towards Maritime Casualties". *Analele Universitatii Maritime Constanta* 11 (13): 23–29. http://ezproxy.lib.swin.edu.au/login?url=http://search.ebscohost.com/login.aspx?direct=true&db=a9h&AN=59156473&site=ehost-live&scope=site.

5 The Case of VTS

Vessel Traffic Services (VTS) are a set of navigational aids established by a port authority or a competent state for monitoring and improving the efficiency and safety of navigation. They are the maritime counterpart to air traffic control, with important differences. There are three types of VTS: information service (provides information about an area), traffic organization service (prevents hazardous situations and increases traffic efficiency), and navigational assistance service (only acts when necessary or at a ship's request). VTS are ruled by the Safety of Life at Sea (SOLAS) Chapter V Regulation 12 together with the Guidelines for Vessel Traffic Services (IMO 1997), adopted by the International Maritime Organization (IMO) on November 27, 1997.

VTS operator tools are primarily radars and communication systems, especially very high frequencies (VHF), and include TV cameras and automatic identification systems (AIS) for ships, which are integrated into a console.

Following the International Association for Lighthouse Authorities (IALA) Human Factors Workshop (IALA 2015), it was clear that an in-depth look at the psychological issues that can affect VTS operators was a major concern.

With the growing need for new professionals to meet the demands of stakeholders in maritime traffic control, it is essential to clarify the challenges to which these professionals are subjected. While the physical challenges of the VTS environment are well documented and have been studied by professionals dedicated to the prevention of occupational hazards in every country, in our opinion the psychological challenges are yet to be determined.

The latest data on VTS personnel from the World VTS Guide ("VTS Finder | World VTS Guide" n.d.) show that, to date there are 191 major VTS around the world, and in every country it is usual that every harbor has a VTS. Each station has an average of 15 operators to ensure 24/7 operation, so around 3000 people work on these major VTS, plus a further 30,000 working on harbor VTS facilities.

The IMO and IALA are involved in the development of policy for improving the standards of qualification and training for VTS personnel, embodied in IMO Assembly Resolution A.857(20) in November 1997 ("IMO | Vision, Principles and Goals" n.d.). A revised SOLAS Chapter V on Safety of Navigation was adopted in December 2000 and entered into force on July 1, 2002. Moreover, the IALA is leading all regulations referring to the organization and operation of the VTS, as well as promoting research and technological development in the field of navigation aids.

In this regard, the IALA has recently promoted a workshop with international human factor (HF) experts related to VTS, where the role of the HF has been discussed. For instance, the upcoming revision of V-119 references using a human-centered design approach to the development, update, and evaluation of VTS. This revision should convey the concept of human reliability, which needs to be carefully considered and reflected by IALA documentation. Therefore, a competent/ VTS authority should develop policy on HF, performance, and capability (IALA 2015, p. 1).

Security policies on navigation aids, and more specifically in the VTS, should focus on the HF; we believe that it is essential to identify the factors and psychological challenges affecting the personnel operating these centers.

This chapter details the major psychological challenges for VTS operators and possible pathologies that can arise. This topic is divided into two main parts: risk factors and protective factors.

5.1 RISK FACTORS

The working environment of the VTS operator is characterized by two issues: shift work and cognitive overload. Other factors, common to other types of work, are teamwork related, coexistence with other professionals, and relationships with superiors and stakeholders. A lack of personnel and personality traits can act as both risk and protective factors.

5.1.1 SHIFT WORK

"Shift work" is a work schedule that involves irregular or unusual hours compared to those of a normal daytime work schedule (Wang et al. 2011). This is the most relevant characteristic of a VTS operator's work, and can affect all aspects of their life.

Evidence indicates that shift work negatively influences workers' physiology, health, safety, sleep, and circadian cycles, resulting in disturbed sleep during the day and excessive sleepiness during the work hours, leading to so-called "shift work disorder" (SWD) (Kalmbach et al. 2015).

Recently, Kalmbach et al. (2015) stated that the consequences of shift work can be depression and anxiety. Åkerstedt and Wright (2009) found that the quality of cognitive processes is optimal when the internal clock is attuned to daylight, so the problem not only occurs because of working night shifts, but also because of shift cycles, which keep the body from properly adjusting its biological clock. These authors also found that a "higher homeostatic sleep drive results in impaired cognition, increased sleepiness, and increased the propensity for sleep" (Åkerstedt and Wright 2009).

We should pay attention not only to the effects on physical and mental health but also to the effects on cognitive performance, in a job precisely characterized by high mental overload.

Monk, Folkard, and Wedderburn (1996) reviewed various critical situations and found five different problem areas:

- Effect on performance
- Sleep and excessive fatigue
- Moodiness, irritability, and disruptiveness
- Absence
- Off-work accidents, e.g., on the drive home

For the first point, there is excellent evidence about the two periods when performance decreases: from around midnight to 06:00 and midday (Monk, Folkard, and Wedderburn 1996). The percentage of errors that could be expected of an operator will be critical around these hours.

Sleepiness and fatigue are frequent problems reported in many studies, as previously described (see Sections 4.1.1 and 4.1.2).

Secondly, sleep and fatigue. Many studies in the field of air traffic control have found evidence showing that any rapid change while on duty, with little sleep, adversely affects the health of workers. Evidence also shows that age is an aggravating factor (Signal and Gander 2007). Takahashi et al. (2005) found similar results in nuclear plants. In addition, some authors found that fatigue increases as the hours pass during a night shift, and this despite the rest the employee may have had during the day; and there is always more perceived sleepiness on night shifts despite the time spent resting (Arslan and Er 2007; Leung et al. 2006; Lützhöft et al. 2007).

Thirdly, the mood in the workplace is sometimes an important issue to consider. An angry and irritable worker can create a disruptive atmosphere that causes communication failure or intolerance and frustration (Kessler et al. 2006). There are also references in the literature on the effects of shifts on the emotional state and how it can affect decision-making, mostly in healthcare personnel (Kalmbach et al. 2015).

Finally, it has been proven that the likelihood of being involved in an accident on the way home from work is almost doubled (Gold et al. 1992).

My experience in several control centers shows that the shift scheduled morning-evening-night provides more time resting and greater comfort ; however, many families, personal, and work-organization circumstances sometimes preferred the *compressed shifts* i.e., to do the largest number of rotations in the shortest time. Rotations impact on a worker's sleep quality and their performance, because they are deprived not only of sleep hours, but also recreation and leisure time (Chambers 1985; Costa 2003; Frost and Jamal 1979; Herbert 1983). People who reside outside the city where the VTS is located often work compressed shifts, as does the worker who decides that this type of shift provides the best balance between rest and family life.

Other contributing factors are forced rotations and service requirements that break the usual pattern of work/rest. On many occasions, a worker is required to cover coworkers' medical leave and vacations or other necessities that have not been provided for by the shift organizer. These circumstances are common in many workplaces but are crucial to consider in a shift system, where the night rotation has a direct impact on the health and performance of the operator who may feel that his usual routine is being disrupted by external demands (Niu et al. 2011).

The impact of these considerations will be different depending on the type of work and the worker's personal characteristics; however, they must be considered by the person organizing the work, who must try to avoid the need for adjustments and last-minute changes.

5.1.2 COGNITIVE OVERLOAD

Cognitive overload is strongly correlated with the advancements in information technology (Karr-Wisniewski and Lu 2010). Although most tasks are automated and operators have many ways to perform their duties, the main part of their work is to maintain situational awareness (SA) (Endsley and Garland 2000).

SA is a dynamic process that requires three steps:

Level 1 SA: Perception: The operator must pay attention to mostly visual, but also auditory inputs. Several active VHF channels must be simultaneously monitored and, often, in several languages.

Level 2 SA: Comprehension: The operator must integrate all of the inputs using the necessary attentional filters to avoid overload their working memory.

Level 3 SA: Projection: The decision-making process occurs. It is based on the operator's prior knowledge (long-term memory) and, especially, on their experience.

All this continuous information processing can result in cognitive overload, which is more pronounced in novice operators as they do not have sufficient training in Levels 2 and 3 SA, i.e., they cannot apply the same attentional filters as a skilled operator because they lack the expertise to do so.

It is common for operators to alternate periods of high attentional overload with other periods of relative boredom, leading to a decrease in attentional skills and difficulty returning to high activation levels when the task requires. During the night, a decrease in circadian rhythms together with a decline in port activity can lead the operator to lower his level of awareness, making it more difficult to become active again (Härmä et al. 2002; Mikko Härmä et al. 1998; Lützhöft et al. 2007; Leung et al. 2006; Niu et al. 2011). This cognitive overload could lead to fatigue, frustration, and human errors (Cordón et al. 2014).

The shift system, as we have seen, also has a great impact on performance. The morning shift is usually the most taxing due to the increased workload (in port VTS mostly) during the early hours, accompanied by other bureaucratic tasks that are often part of an operator's work. In other VTS, there is no decrease in traffic during the night shift, so it is essential that the operator can take periods of rest to maintain the necessary attention overnight.

Similarly, if an operator finishes a night shift and starts another shift the next afternoon, he can probably expect a significant drop in his performance mid-afternoon, which must be considered.

Are the first hours of the morning the most dangerous for maintaining SA, as these hours correspond with low SA activation and an increase in ships' maneuvers and port traffic?

5.1.3 STRESS

Stress is one of the most studied constructs in work psychology. Stress happens when there is a mismatch in the person–environment interaction (real or perceived) between situational demands and the capabilities to meet these demands (physiological or psychological) in a certain environment. Such imbalances cause stress.

González Cabanach (1998) distinguishes some potentially stressful sources in the work context, including the physical conditions in which the work is performed, work overload and the availability of resources, labor, and other content intrinsic to the work itself. Factors related to

- The performance of roles.
- Interpersonal relationships generated in professional life.
- Career development.
- The structure and organizational climate.
- Insufficient autonomy.

Stress is not a unique concept, and we must differentiate between daily and specific stressors. Stressors in a VTS operator's environment, as already mentioned, include shift work and the need to maintain awareness. However, other daily stressors must be taken into account: the mismatch between work and social life is the most important, but another stressor is the absence of professional development or insufficient capacity to evolve within the organization. Another significant source of stress is teamwork; some workers' insufficient understanding or their inability to empathize might generate frustration in their coworkers (Grandjean et al. 1971; Hoffman and Scott 2003; Rengamani and Murugan 2012; Sneddon, Mearns, and Flin 2013).

The main effect of stress that concerns us is cognitive. The key issues are difficulty making decisions; feeling confused; inability to concentrate; difficulties maintaining attention; insufficient control over feelings; *narrowing* of attention; disorientation; forgetfulness; mental blocks; and hypersensitivity to criticism. The consequences on a worker's health, performance, and job satisfaction are also well known (Beehr and Newman 1978).

All these effects can negatively influence the performance of operators because they can increase absenteeism and burnout and affect decision-making processes.

5.1.4 WORK–LIFE BALANCE

Dealing with the work shift system is a crucial challenge for VTS operators. In recent studies (KENEXA 2007; NHIS 2013; Reynolds 2005), 16% of US workers reported having serious problems in reconciling work and family life, most prevalent between the ages of 30 and 44 years (probably due to parenting). The work–family balance affects both genders, but coping techniques are different. Also, cultural principles play an important part, due to the difference in family roles present in many societies. This can be a key factor that has a significant influence on stress, or even causing occupational disabilities.

For many people, private life activities are *obligatory* (mainly in Western countries), producing personal and family conflicts, feelings of constant stress and job dissatisfaction, which can lead to burnout (Chambers 1985; Frost and Jamal 1979; Herbert 1983; NHIS 2013). This is particularly the case for mothers, when social pressures force them to combine professional and family roles (Heraty et al. 2008).

5.2 PROTECTIVE FACTORS

It is impossible to differentiate one personality aspect as a protective factor, as it depends on important personality traits. The same situation can be perceived as highly stressful by one person and meaningless by another person.

The main protecting factor at work is resilience, which is the quality of being psychologically durable and able to withstand adversity (Caza and Milton 2011; Greve 2009). Resilience is a protective factor not only in the workplace, but also in daily life; therefore, it is a factor demanded by recruiters, and should be one of the main virtues of a VTS operator. Many studies on resilience in the workplace suggest that it is a process of interaction between the individual and his environment, and not a fixed and immutable personality trait, which makes resilience a trainable trait.

Similarly, the mechanisms to manage stress are both, relatively innate or trainable. Although a certain level of stress is always desirable as it can be understood as cognitive activation, it is important to balance these factors.

Sport in general and an active and balanced lifestyle are commonly accepted as factors in personal and work protection (Rutter 1987), and facilitate the regulation of sleep patterns.

Proper sleep hygiene and proper life planning are the most important protective factors against fatigue generated by shift changes (Barber, Grawitch, and Munz 2013). However, these are often not possible to accomplish due to the operational needs of the VTS, or in a few cases, due to personal life pressures that force operators to take unhealthy shifts.

Many studies suggest that the impact of the shift system can be minimized by choosing rotations with lower health consequences, both physical and psychosocial (Knauth 1996; Knutsson 2003; Monk, Folkard, and Wedderburn 1996; Takahashi et al. 2005).

5.3 CONCLUSION

With increasing requests for professionals to meet the demands of the VTS, a review of the psychological working conditions to which these professionals are subjected is needed. While the physical conditions are not highly demanding, the fact of being subjected to shift work, the need to constantly maintain SA, and the consequences of decisions made by operators are the main determining factors, in my opinion, of these jobs.

As previously stated, during the recruitment process, the authorities in charge of the VTS are encouraged to promote the enrollment of crews with resilience and adaptability.

During the training process, the authorities should provide candidates with the tools to deal with stress management, personal organization, and emotional conflict with peers. It would also be useful to teach sleep hygiene and how sleep works, both to prevent sleepiness on duty and to make sleep restful.

In our opinion it is essential that the worker has advance notice of the shift scheduling, to facilitate the work–family balance as much as possible. Introducing extra shifts or breaking the rate of work/rest should be avoided. Crews should plan and agree well in advance any changes to rotations.

Also, operators over a certain age should be given the opportunity to opt out of night shifts, because while some people adapt well to them, others suffer, especially women. In this regard, recall the recommendations of the International Labor Organization (ILO (UN) 1990).

VTS operators/supervisors are encouraged to have a proper work–family life balance, adapting it to their age and physical condition with an active and healthy lifestyle. Try to avoid stress caused by shift changes to keep up with the pace of modern life. A proper plan should be made of their work and personal agenda and everything should be planned well in advance.

Strategies and recommendations must be developed and implemented by VTS organizations to increase efficiency and labor engagement, which can be achieved through specific training programs on stress coping strategies, time management, resilience, promoting an active and healthy lifestyle, sleep hygiene, etc. These programs could be provided during trainers and managers courses or through online learning programs or on-the-job training.

The improvement of safety and security in Maritime Transport is the ultimate objective of this strategy, thus, IALA should promote the quality of service to stakeholders and the improvement in the quality of life of VTS personnel by publishing recommendations for Members and National Maritime Authorities.

REFERENCES

Åkerstedt, Torbjörn, and Kenneth P Wright. 2009. "Sleep Loss and Fatigue in Shift Work and Shift Work Disorder". *Sleep Medicine Clinics* 4 (2): 257–271.

Arslan, Özcan, and İsmail Deha Er. 2007. "Effects of Fatigue on Navigation Officers and SWOT Analyze for Reducing Fatigue Related Human Errors on Board". *International Journal on Marine Navigation and Safety of Sea Transportation* 1 (3): 345–349.

Barber, Larissa, Matthew J Grawitch, and David C Munz. 2013. "Are Better Sleepers More Engaged Workers? A Self-Regulatory Approach to Sleep Hygiene and Work Engagement". *Stress and Health: Journal of the International Society for the Investigation of Stress* 29 (4): 307–316.

Beehr, Terry A, and John E Newman. 1978. "Job Stress, Employee Health, and Organizational Effectiveness: A Facet Analysis, Model, and Literature Review". *Personnel Psychology* 31 (4): 665–699.

Caza, Brianna Barker, and Laurie P Milton. 2011. *Resilience at Work. Building Capability in the Face of Adversity*. New York: Oxford University Press.

Chambers, D. 1985. "Shift Work and Leisure". *Temps Libre* 14: 42–46.

Cordón, José R, Pedro Ramiro Olivier, Manuel A García Sedeño, and Jorge Walliser Martín. 2014. "Diseño y Validación de una Prueba de Selección Para Controladores de Tráfico Marítimo Basada en la Medida de la Conciencia Situacional". *Revista de Psicología del Trabajo y de las Organizaciones* 30 (2): 83–93.

Costa, Giovanni. 2003. "Shift Work and Occupational Medicine: An Overview". *Occupational Medicine* 53 (2): 83–88.

Endsley, Mica R, and Daniel J Garland. 2000. *Situation Awareness Analysis and Measurement*. Lawrence Erlbaum Associates. http://books.google.ca/books?id=AxYerHP5f-sC.

Frost, Peter J, and Muhammad Jamal. 1979. "Shift Work, Attitudes, and Reported Behavior: Some Associations between Individual Characteristics and Hours of Work and Leisure". *Journal of Applied Psychology* 64 (1): 77–81.

Gold, D R, S Rogacz, N Bock, T D Tosteson, T M Baum, F E Speizer, and C A Czeisler. 1992. "Rotating Shift Work, Sleep, and Accidents Related to Sleepiness in Hospital Nurses". *American Journal of Public Health* 82 (7): 1011–1014.

González Cabanach, R. 1998. "Comunicación, Estrés y Accidentabilidad. Tres Factores de Actualidad". *Fotocopia Personal*.

Grandjean, E P, G Wotzka, R Schaad, and A Gilgen. 1971. "Fatigue and Stress in Air Traffic Controllers". *Ergonomics* 14 (1): 159–165.

Härmä, M, Mikael Sallinen, R Ranta, P Mutanen, and K Müller. 2002. "The Effect of an Irregular Shift System on Sleepiness at Work in Train Drivers and Railway Traffic Controllers". *Journal of Sleep Research* 11 (2): 141–151.

Härmä, Mikko, Leena Tenkanen, Tom Sjöblom, Tiina Alikoski, and Pertti Heinsalmi. 1998. "Combined Effects of Shift Work and Life-Style on the Prevalence of Insomnia, Sleep Deprivation and Daytime Sleepiness". *Scandinavian Journal of Work, Environment and Health* 24 (4): 300–307.

Heraty, Noreen, Michael J Morley, Jeanette N Cleveland, Geraldine Grady, and Alma M McCarthy. 2008. "Work-Life Integration: Experiences of Mid-Career Professional Working Mothers". *Journal of Managerial Psychology* 23 (5): 599–622.

Herbert, A. 1983. "The Influence of Shift Work on Leisure Activities A Study with Repeated Measurement". *Ergonomics* 26 (6): 565–574.

Hoffman, Amy J, and Linda D Scott. 2003. "Role Stress and Career Satisfaction among Registered Nurses by Work Shift Patterns". *The Journal of Nursing Administration* 33 (6): 337–342.

IALA. 2015. "Summary Report of the Workshop On Human Factors and Ergonomics in VTS". Vol. 16. Gothenburg, Sweden.

ILO (UN). 1990. "Recommendation R178 - Night Work Recommendation, 1990 (No. 178)". http://www.ilo.org/dyn/normlex/en/f?p=NORMLEXPUB:12100:0::NO::P121 00_INSTRUMENT_ID:312516.

"IMO I Vision, Principles and Goals". 2017. Accessed December 17, 2014. http://www.imo.org/OurWork/HumanElement/VisionPrinciplesGoals/Pages/Default.aspx.

IMO. 1997. *Resolution A.857(20) Adopted on 27 November 1997 Guidelines for Vessel Traffic Services.* 20th session, Agenda item 9. International Maritime Organization.

Kalmbach, David A, Vivek Pillai, Philip Cheng, J Todd Arnedt, and Christopher L Drake. 2015. "Shift Work Disorder, Depression, and Anxiety in the Transition to Rotating Shifts: The Role of Sleep Reactivity". *Sleep Medicine* 16 (12): 1532–1538.

Karr-Wisniewski, Pamela, and Ying Lu. 2010. "When More Is Too Much: Operationalizing Technology Overload and Exploring Its Impact on Knowledge Worker Productivity". *Computers in Human Behavior* 26 (5): 1061–1072.

KENEXA. 2007. "Kenexa Research Institute Finds That When It Comes to Work/Life Balance, Men and Women Are Not Created E". 2007. https://web.archive.org/web/20071014203018/http://www.kenexa.com/en/AboutUs/Press/2007/07JUL25.aspx.

Kessler, Ronald C, Hagop S Akiskal, Minnie Ames, Howard Birnbaum, Paul Greenberg, Robert Jin, Kathleen R Merikangas, Gregory E Simon, and Philip S Wang. 2006. "Prevalence and Effects of Mood Disorders on Work Performance in a Nationally Representative Sample of US Workers.". *American Journal of Psychiatry* 163 (9): 1561–1568.

Knauth, P. 1996. "Designing Better Shift Systems". *Applied Ergonomics* 27 (1): 39–44.

Knutsson, Anders. 2003. "Health Disorders of Shift Workers". *Occupational Medicine* 53 (2): 103–108.

Leipold, Bernhard, and Werner Greve. 2009. "Resilience". *European Psychologist* 14 (1): 40–50.

Leung, Ada W S, Chetwyn C H Chan, Jimmy J M Ng, and Peter C C Wong. 2006. "Factors Contributing to Officers' Fatigue in High-Speed Maritime Craft Operations". *Applied Ergonomics* 37 (5): 565–576.

Lützhöft, Margareta, Birgitta Thorslund, Albert Kircher, and Mats Gillberg. 2007. "Fatigue at Sea: A Field Study in Swedish Shipping". Statens Väg-Och Transportforskningsinstitut.

Monk, Timothy H, Simon Folkard, and Alexander I Wedderburn. 1996. "Maintaining Safety and High Performance on Shiftwork". *Applied Ergonomics* 27 (1): 17–23.

NHIS. 2013. "CDC – The 2010 Occupational Health Supplement to the National Health Interview Survey (NHIS) Home Page – NIOSH Workplace Safety and Health Topic". National Institute for Occupational Safety and Health, Centers for Disease Control and Prevention. http://www.cdc.gov/niosh/topics/nhis.

Niu, Shu-Fen, Min-Huey Chung, Chiung-Hua Chen, Desley Hegney, Anthony O'Brien, and Kuei-Ru Chou. 2011. "The Effect of Shift Rotation on Employee Cortisol Profile, Sleep Quality, Fatigue, and Attention Level: A systematic review". *The Journal of Nursing Research: JNR* 19 (1): 68–81.

Rengamani, J, and M Sakthivel Murugan. 2012. "A Study on the Factors Influencing the Seafarers' Stress". *MET International Journal of Management* 4 (1): 44–51.

Reynolds, Jeremy. 2005. "In the Face of Conflict: Work-Life Conflict and Desired Work Hour Adjustments". *Journal of Marriage and Family* 67 (5): 1313–1331.

Rutter, Michael. 1987. "Psychosocial Resilience and Protective Mechanisms". *The American Journal of Orthopsychiatry* 57 (3): 316–331.

Signal, T Leigh, and Philippa H Gander. 2007. "Rapid Counterclockwise Shift Rotation in Air Traffic Control: Effects on Sleep and Night Work". *Aviation, Space, and Environmental Medicine* 78 (9): 878–885.

Sneddon, Anne, Kathryn Mearns, and Rhona Flin. 2013. "Stress, Fatigue, Situation Awareness and Safety in Offshore Drilling Crews". *Safety Science* 56 (July): 80–88.

Takahashi, Masaya, Takeshi Tanigawa, Naoko Tachibana, Keiko Mutou, Yoshiko Kage, Lawrence Smith, and Hiroyasu Iso. 2005. "Modifying Effects of Perceived Adaptation to Shift Work on Health, Wellbeing, and Alertness on the Job among Nuclear Power Plant Operators". *Industrial Health* 43 (1): 171–178.

"VTS Finder | World VTS Guide". n.d. Accessed March 11, 2016. http://www.worldvtsguide. org/VTS-Finder.

Wang, X-S, M E G Armstrong, B J Cairns, T J Key, R C Travis, and P J Nicholson. 2011. "Shift Work and Chronic Disease: The Epidemiological Evidence". *Occupational Medicine* 61 (6): 443–444; author reply 444.

Index

Administration by exception, 23
AIS. *See* automatic identification systems
Åkerstedt,T. 88, 102
Akhona Geveza, 81
Alarm Signals, 39
Alfiani, D.S. *See* Multiculturalism
Allen, E. 93
Allport, F. H. 57, *See* intergroup harmony
American Psychological Association. *See*
    Resilience
apprehension by evaluation, 57
Arendt, J. 86
Aritzeta, A. *See* Teamwork
Arslan, O. 103
Arthur, M. B. *See* 2.1.5.1 Path to the Goal Model
Asch, S. E. 52, *See* Conformism
Assertiveness, 42, 43
Attributional biases, 29
Audickas, S. *See* diversity
authentic leadership, 33, 35
Authority, 8
Autocratic. *See* 2.1.5.1 Model of Leader
    Participation
Automatic Identification Systems, 103
Avoidance of uncertainty, 31
Ayestaran, S. *See* Teamwork

Balanced processing of information, 35
Baltic and International Maritime Council, 71
Barata, M. A. *See* Globe Project
Barber, L. 107
Bass, B. M. 8, *See* 2.1.9.1 Transformational
    Leadership
Baulk, S. 93, *See* circadian cycles
Beehr, T. A. 106
Benne, K.D. *See* Group roles
Benton, G. *See* Multiculturalism
Berson, Y. *See* organizational climate
Bierhoff, H. W. 41
BIMCO. *See* Baltic and International Maritime
    Council
Black Sheep, 50
Blake, R. R. *See* Conformism
Blanchard, K H. *See* 2.1.5.1 Hersey and Blanchard
    Situational Leadership Model
Blascovich, J. *See* challenge pattern and threat
    pattern
Boin, A. *See* emergencies
Bond, M. H. *See* cultural dimensions, *See*
    Conformism
Brenker, M. 74, *See* Multiculturalism

Brewer, M. B. 63
bridge resource management, 36
BRM. *See* Bridge Resource Management
Brown, M. E. *See* ethical leadership, *See*
    Prejudices
Burch, J. B. 91
Burn, S. M. 59
Burns. *See* Transformational leadership
Busan Declaration, 82

caffeine, 91
Carli. L. L. *See* Conformism
Carol-Dekker, L. 71, 81, 85
Castro Solano. A. *See* leadership models focused
    on followers
Categorization, 27
Caza, B. B. 107
Cerebral Hemispheres, 38
challenge pattern and threat pattern, 57
Chambers, D. 103, 106
chaos, 44
charisma, 21, 24
charismatic leadership, 21
chronemics, 78
circadian cycles, 86
circadian rhythms, 88
Clark, R. D. *See* 3.1.8 Minority infLuence
Coleman. J. F. *See* Conformism
collectivism, 31, 68
communication failures, 72
conformism, 51, 52
consensus, 52
consultative. *See* 2.1.5.1 Model of Leader
    Participation
contingency models, 2-9
contingent reward, 22
Cordón, J. R. 105
Costa, G. 103
Costa Concordia, 5
Cottrell. N. B. *See* apprehension by evaluation
cross-cultural, 31
crowd control, 44, 40
Crutchfield, R. S. *See* Conformism
cultural dimensions, 66

Davis. C. *See* diversity
De Martino. B. *See* emergencies
decision-making under pressure, 38
democratic and autocratic leadership. *See* Lewin
Deutsch, M. *See* Conformism
discrimination, 27

disposition of the follower. *See* 2.1.5.1 Hersey
      and Blanchard Situational Leadership
      Model
distance of power, 30-31
diversity, 66
Doyle, N. *See* Resilience
Dragomir, C. *See* gender
drowsiness, 88
dynamic leaders. 8

Eagly, A. H. *See* Conformism
Edwards, B. D. *See* job satisfaction
effect of coercion, 57
Eler, G. *See* gender
emergencies, 38
emergency drills, 40
empowerment of teams, 65
Endsley, M. 105
Environmental uncertainty. *See* External factors
Er, H. 103
Estonia, 40
ethical leadership, 33
evil leadership, 35
external factors, 9

fatigue, 92
femininity, 31
Festinger, L. 52
Fiedler, F. E. *See* Fiedler's model
Fielder's Model, 10
Flin, R. 106
Folkard, S. 102, 107
Folkman. J. *See* Resilience
Fort, C. A.. *See* non-verbal communication
Franke, R. H. *See* cultural dimensions
Froholdt, L.L: 82
Frost, P. J. 103, 106
fundamental attribution error, 29

Gander, P. H. 103
Garland, D. J. 105
gender, 33
Gerard, H. *See* Conformism
Gerstenberger, H. 72
Gil, F. *See* Globe Project
glass ceiling, 33
Globe Project, 31
Gold, D. R. 103
González Cabanach, R. 105
Goodwin, V. L. *See* leadership models focused
      on followers
Gordon, W. 7
Gouran, D. S.. *See* Shared information bias
Grandjean, E. P. 106
Grawitch, M. J. 107
Greve, W. 107
group diversity, 48

group performance, 56
group roles, 50
group size, 47
group thinking, 62
Guetzkow, H. 52, *See* Conformism

haptic communication, 76
Härmä, M. 88, 105
Harrison, D. A. *See* ethical leadership
Hellige, J. B. *See* emergencies
Heraty, N. 106
Herbert, A. 103
Hersey, P. *See* 2.1.5.2 Hersey and Blanchard
      Situational Leadership Model
Hersey and Blanchard Situational Leadership
      Model, 10
hierarchical distance, 66
hierarchies, 49
Hilton, J. L. *See* stereotypes
Hirokawa, R. Y. *See* shared information bias
Hochbaum, G. M.. *See* Conformism
Hoffman, A. J. 106
Hofstede, G. 66, 68, *See* cultural dimensions,
      See masculinity, *See* cross-cultural
Hogan, B. 85
Hogg. M. A. *See* Intragroup differentiation
Hollander, E. P. *See* influence
Horck, J. *See* Multiculturalism
House, R. J. *See* 2.1.5.1 Path to the Goal Model
hypothalamus, 41

IALA. *See* International Association for
      Lighthouse Authorities
ICS. *See* International Chamber of Shipping
idealized influence, 22
idiosyncratic credit. *See* influence
idiosyncratic credit theory, 20
illusion of unanimity, 62
illusory correlation, 27
ILO, 95, 107, *See* International Labour
      Organization
IMO, 43, 72, 81, 82, 92, 93, 94, *See* International
      Maritime Organization
implicit leadership, 18
impulse theory, 57
individualism, 68,31
individualized consideration, 22, 24
influence, 20
Insko, C.A. *See* Conformism
inspiration, 24
inspiring motivation, 22
integrated workplace management system, 82
intellectual stimulation, 22, 24
intergroup harmony, 29
intergroup hostility, 27
intergroup phenomena, 26
intergroup relations, 25

internal factors, 9
internalized morality, 35
International Association for Lighthouse
    Authorities, 103
International Chamber of Shipping, 71
International Convention on Standards
    of Training, Certification and
    Watchkeeping for Seafarers, 5
International Labour Organization, 72
International Maritime Organization (IMO)
IMO, 17
intragroup differentiation, 48
IWMS. See integrated workplace management
    system

Jamal, M. 103, 106
Janis, I. L. 62, See Shared information bias
Jensen, O. C. 72, 85, 96
job satisfaction, 30
Jones, E. E. See Conformism

Kalmbach, D. A. 102, 103, See shift work
    disorder
Karr-Wisniewski, P. 103
Kerr, N. L. See Köhler effect
Kessler, R. C. 103
Khan, S. 71, 81, 85
kinesis, 77
Kitada, M. 82
Klein, R. 41
Knauth, P. 107
Knutsson, A. 107
Koester, T. 72
Köhler effect, 59
Kramp, P. 41
Kumar, K. See diversity

laissez-faire, 24
Lapa, K. 86, 93
leadership, 2, 5
leadership in emergency, 37
leadership models focused on followers, 18
leadership styles approach, 9
left frontotemporal area, 38
left hemisphere, 38
legitimacy. See influence
Lepore, L. See Prejudices
Leung, A.W.S. 103, 105
Levey, G. B. 81
Levine, J. M. 47, 48
Lewin, K. 9
Linton, J.D. See organizational climate
locus of control, 42, See 2.1.5.1 Path to the
    Goal Model
Lott, A.J. See Conformism
Lu, Y. 103
Lützhöft, M. 103, 105

Maass, A. See 3.1.8 Minority infLuence
Magramo, M. See gender
Manila Amendments, 17
Mann, L. 62
Manzano, J. 81
Mari, P. See Multiculturalism
Maria M, 35
Marine Insight, 82
Maritime Rescue Coordination, 36
Marques, J.M. See Black Sheep
MARTHA, 90
Martha Project, 86
masculinity, 31
masculinity and femininity, 68
Mearns, K, 106
Meertens, T.F. See Prejudices
melatonin, 90, 91
Michaelsen, K. See diversity
Milton, L.P. 107
minority influence, 54
MLQ. See multifactor leadership questionnaire
modal communication, 77
Model of Leader Participation, 13
Modood, T. 81
Monk, T. H. 102, 107
Moreland, R.L. 47, 48
Moscovici, S. See 3.1.8 minority influence
Mouton, J.S. See Conformism
moutza, 80
MRCC. See Maritime Rescue Coordination
    Center
MS Estonia, 43
Muehlbach, M.J. 91
Mukherjee, P. 82
multiculturalism, 2, 69
multifactor leadership questionnaire, 24
Mummendey, A. 63
Munz, D.C. 107
Murugan, M.S. 106

Natemeyer, W.E. See 2.1.5.1 Hersey and
    Blanchard Situational Leadership
    Model
Newman, J.E. 106
NGV 2000, 43
NHIS, 106
Nijstad, B.A. See 3.2.1 group performance
Nisbet, R.E. See attributional biases
Niu, S.F. 103, 105
Nogareda Cuixart, C. 90
non-traditional organizations, 33
non-verbal communication, 74
NSQ. See suprachiasmatic nucleus

Oculesics, 77
officer on watch, 98
Oldenburg, M. 85, 96

Omar, A. 30, *See* job satisfaction
OOW, 99, *See* officer on watch
organizational climate, 30
Otten, S. 63

Paez, D. *See* Black Sheep
paradigm of coercion, 57
paralanguage, 74
Paris, L. *See* job satisfaction
path to the goal model, 14
Pauksztat, B. 86
personal bubble, 76
persuasive argument, 60
perverse rules, 50
Pettigrew, T.F. *See* Prejudices
polarization, 60
position of power. *See* Fiedler's model
power, 8
prejudices, 26
pressure toward uniformity, 50
pressure training, 40
problem of the stowaway, 59
procedural or organic memory, 40
Progoulaki, M 74, *See* Multiculturalism
proxemics, 76
pseudo-collaborators, 40
psychologization, 54
Pyne, R. *See* communication failures

Ramadan, 80
Rengamani, J, 106
resilience, 41, 107
Reyner, L. 93, *See* circadian cycles
Reynolds, J. 106
right hemisphere, 38
right occipital lobe, 38
Ringelmann effect, 57
Ripoll, M. *See* Globe Project
Roe, M. 74, *See* Multiculturalism
Rosch, E. *See* LeaDership MoDeLs focuseD
        on foLLowers
Ross, E. *See* Attributional biases
Rossi, E. *See* emergencies
Rutter, 107

SA. *See* situational awareness
Safety of Life at Sea, 103
saikerei, 80
Saksvik, I.B. 88
Schweitzer, P.K. 91
Scott, L.D. 106
self-awareness, 35
self-categorization, 27, 61
self-directed team, 63
Shamir, B. *See* 2.1.5.1 Path to the Goal Model
shared information bias, 61
Sheats, P. *See* Group roles

Sherif, M. 52, *See* intergroup relations
shift work disorder, 88
Shwetha, B. 88
Signal, T. L. 103
situational awareness, 105
situational bias, 29
sleep hygiene, 91, 107
SMCP, 74, *See* Standard Marine Communication
        Phrases
Smith, P.B. 93, *See* Conformism
Sneddon, A. 106
social comparison, 60
social compensation, 59
social cryptomnesia, 54
social facilitation, 57
social identity, 26
social influence, 51
Solano, A. C. 7, 8
SOLAS, 103, *See* Safety of Life at Sea
Standard Marine Communication Phrases, 72
states of fear, 40
status system, 49
STCW, 43, 44, 72, 96
Steiner, I,D. *See* 3.2.1 group performance
Stephan, C.W. 63
stereotypes, 26
Stogdill, R.M. 8, *See* 2.1.9.1 Transformational
        Leadership
suprachiasmatic nucleus, 88
Surugiu, F. *See* gender
Swailes, S. *See* Teamwork
SWD. *See* shift work disorder
Szczepańska, M. *See* diversity

Taber, T. 7
Takahashi, M. 103, 107
Tallinn, 41
teamwork, 2, 47
TenHouten, W. *See* emergencies
Terry, D.J. *See* Intragroup differentiation
The International Convention on standards
        of training, certification, and
        Watchkeeping for seafarers
STCW, 17
The Manila Amendments, 5
Thomas, M. *See* gender
Titanic, 35
traditional organizations, 33
trait approach, 8
trajectory goal. *See* 2.1.5.1 Path to the Goal Model
transactional, 22, 24, 33
transactional leadership, 22
transformational, 22, 24, 33
Transparency in relationships, 35
Treviño, L.K. *See* ethical leadership
typology of leadership. *See* 2.1.5.1 Hersey and
        Blanchard Situational Leadership Model

uncertainty, 52
University of Cardiff, 93
uprooting, 96
US Coast Guard, 83
USCG, 98, *See* US Coast Guard

Van Engen, M.L.. *See* gender
very high frequencies, 103
Vessel Traffic Services
VTS, 3
VHF. *See* very high frequencies
Von Hippel, W. *See* stereotypes
Vroom, V.H. *See* 2.1.5.1 Model of Leader
        Participation
VTS, 98, 99, 103, *See* Vessel Traffic Services

Wadsworth, E. 93
Walsh, J. K. 91
Walumbwa, F.O. *See* authentic leadership
Wang, X. 88, 102

Warren Bennis, 6
Warsash Maritime Academy. *See* Martha Project
Watson, W.E. *See* diversity
Wedderburn, A.I. 102, 107
Whittington, J.L. *See* LeaDership MoDeLs
        focuseD on foLLowers
Willemsen, T.M. *See* gender
Williams, P. 82
Wofford, J.C. *See* LeaDership MoDeLs focuseD
        on foLLowers
World Maritime University, 82
Wright, K.P. 88, 102

Xhelilaj, E. 86, 93

Yetton, P.W. *See* 2.1.5.1 Model of Leader
        Participation
Yukl, G. 7

Zajonc, R.B. *See* impulse theory

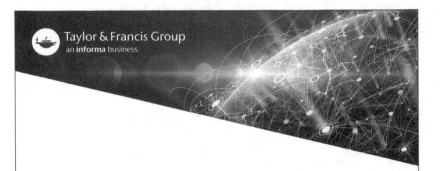